Nanoclay - Recent Advances, New Perspectives and Applications

Edited by Walid Oueslati

Published in London, United Kingdom

Nanoclay - Recent Advances, New Perspectives and Applications
http://dx.doi.org/10.5772/intechopen.102100
Edited by Walid Oueslati

Contributors
Oluwatoyin Gbadeyna, Zikhona Linganiso, Nirmala Deenadayalu, Walid Oueslati, Abdesslem Ben Haj Amara, Chadha Mejri, Boubié Guel, Issaka Garikoé, Kawthar Yahya, Noureddine Hamdi, Wissem Hamdi, Houda Saad, Ameni Ayed, Mondher Srasra, Sameh Attia, Ezzeddine Srasra, Fatima Charrier-El Bouhtoury, Olfa Tabbene, Serap Yalcin, Nevin Cankaya

First published in London, United Kingdom, 2022 by IntechOpen
IntechOpen is the global imprint of INTECHOPEN LIMITED, registered in England and Wales, registration number: 11086078, 5 Princes Gate Court, London, SW7 2QJ, United Kingdom

British Library Cataloguing-in-Publication Data
A catalogue record for this book is available from the British Library

Additional hard and PDF copies can be obtained from orders@intechopen.com

Nanoclay - Recent Advances, New Perspectives and Applications
Edited by Walid Oueslati
p. cm.
Print ISBN 978-1-80356-557-6
Online ISBN 978-1-80356-558-3
eBook (PDF) ISBN 978-1-80356-559-0

Meet the editor

Dr. Walid Oueslati, Ph.D., HU, is an associate professor in the Physics Department, Faculty of Science, at the University of Carthage, Bizerte, Tunisia. He is a permanent member of the Resources, Materials and Ecosystems research group at the university. His research interests include experimental physics, materials science, mineralogy, crystallography, solid-state physics, and condensed matter, with particular reference to X-ray powder diffraction structure analysis and lamellar disordered material. His current research activities are mainly focused on the preparation and structural characterization of hybrid materials, the synthesis of nanomaterials, the use of clay nanofillers in clay/polymer nanocomposites, phyllosilicate substrates of inorganic/organic compounds, perovskite structure, and mesoporous/microporous materials.

Contents

Preface

This book presents state-of-the-art findings on the characterization and uses of nanoclay or nano-layered silicate, which are optimized clay minerals with a number of improved features. Many geological, technological, and biological processes are supported by adsorption, wetting, dispersion, coagulation, sedimentation, structure formation, capillary, electrokinetic, ion exchange, mechanical response, and other nanoclay phenomena.

This work's primary goal is to contribute to a complete knowledge of these processes, taking into account the rapid growth of nanotechnologies. Additionally, the book gives a thorough overview of how nanoclay is used as reinforcing filler among different nanoparticles, with a special emphasis on polymer and biopolymer-clay nanocomposites. Improvements in mechanical, thermal, electrical, flame-retardant, and gas barrier properties are also seen when clay nanoparticles are added to polymers. Students, academics, and researchers who want to learn more about nanoclay and its uses in the areas of nanotechnology, biotechnology, environmental science, and industrial remediation will find this book helpful.

Walid Oueslati
Faculty of Sciences of Bizerte,
Physics Department,
University of Carthage,
Bizerte, Tunisia

Chapter 1

X-ray Diffraction Profiles Modeling Method for Layered Structures Reconstruction: Nanoclay Structural Verification

Walid Oueslati, Chadha Mejri and Abdesslem Ben Haj Amara

Abstract

The nanoclay properties find a large environmental application domain as depolluant, ion exchanger, natural geological barrier for industrial and radioactive waste confinement, clay-based nanocomposite for drug delivery, and more. Layered materials, such as nanoclay, present rather complex structures whose classical characterization requires a complementarity between several analysis methods to decipher the effects of interstratification (and its cause) on the intrinsic functional properties. The appearance of defects related to the layers stacking mode, which differ in their thickness and/or their internal structure are directly related to the reactivity of the mineral's surface. During the last decades, and with the development of computer codes, the modeling of X-ray diffraction profiles has proven to be an important tool that allows detailed structural reconstruction. The quantitative XRD analysis, which consists of the comparison of experimental (00l) reflections with the calculated ones deduced from structural models, allowed us to determine the optimal structural parameters describing interlamellar space (IS) configuration, hydration state, cation exchange capacity (CEC), layer stacking mode, and theoretical mixed-layer structure (MLS) distribution. This chapter will review the state of the art of this theoretical approach as a basic technique for the study of nanoclays. The basic mathematical formalism, the parameters affecting the theoretical models, and the modeling strategy steps will be detailed in concrete examples.

Keywords: nanoclay, layered materials, modeling of X-ray diffraction profiles, layer thickness, layer stacking mod, modeling strategy

1. Introduction

Hydration properties of nanoclays are controlled by several factors such as the type of the interlayer cation and the amount and the layer charge location (created by isomorphic substitutions in octahedral or tetrahedral sites). The nanoclay swelling process is controlled by the balance between the repulsive force owing to the layer interactions and the attractive forces between exchangeable cations and the negatively charged surface of siloxane layers [1–4]. XRD is the basic analysis method for the structural characterization of clay minerals. The structural characterization of nanoclays, and given their nanoscopic properties, through a simple XRD qualitative analysis does not provide clear and convincing answers if we really want to take advantage of these intrinsic properties. Indeed, the qualitative XRD investigation based on the basal spacing d_{001} from the first order (001) Bragg reflections, the full-width at half maximum (FWHM), the profile geometry (i.e., symmetric or asymmetric reflections) description, and the deviation rationality parameter (ξ), respectively (reference), remains insufficient for a precise description. The information remains obsolete and incomplete on what is really happening at the level of the IS configuration and its content as well as the layers distribution within the crystallites. Also, analysis of XRD effects from defective nanoclay materials cannot be reached adequately with conventional XRD methods such as single-crystal diffraction and/or Rietveld structure refinement because of this reduced periodicity. This has driven the advancement of specific algorithms for the calculation of diffraction intensity occurring from defective structures.

The development of specific modeling XRD techniques based on the theoretical approach is imposed. Pioneering authors [5–14] studied the IS configuration focusing on the atomic positions of exchangeable cations and the associated H_2O molecules of Na-montmorillonite samples by assuming a homogeneous hydration state and neglecting the coexistence of different hydration states in a sample. The heterogeneous layer charge distribution is investigated by [9] who showed that the insertion of the IS water molecule is accompanied by a progressive expansion of the basal spacing value which is done by the discrete hydration state going from the dehydrated (0 W, d_{001} = 10 Å) to the strongly hydrated ones (4 W, $d_{001} \approx$ 21 Å) passing by the 1 W, 2 W, and 3 W hydration states [9, 15]. Among the problems solved by XRD modeling approach is the coexistence of several hydration states within the same layer and the coexistence of different types of layers within the same crystallite.

Indeed, [16] demonstrated that the presence of different exchangeable cations in different interlayers is lead to the presence of segregated domains. These domains attributed to the demixed state are described in the works of [17]. The modeling XRD approach correlated to the adsorption–desorption measurements are used by [18] to determine the proportion of the different layer types coexisting along the isotherms. More recently, [19–26] used this approach to fit reflection positions and 00l profiles over a large angular range focusing study of the smectite hydration behavior and the selective ion exchange process.

They demonstrate the existence of mixed-layer structure (MLS) and a specific response to mechanical and geochemical disturbances has been resolved. Authors [27–28] showed the great possibility of using diffraction techniques to root out structural information from poor 3D crystal periodicity. They also show that a reliable characterization of the structural and chemical heterogeneities of the layered structures mainly depends on an optimized and reliable interpretation of the diffraction data [29]. Several proposed methods allowing the theoretical calculation of the diffracted intensity for the non-periodic interstratified structures have been proposed in the literature [30–33]. The proposed models present some technical and scientific failures. The main development was based on a formalism developed by [7, 13, 34–37] to describe the diffracted 00l and hkl intensities by a set of crystals containing different layer types.

The XRD modeling method is used to quantify the MLS, the layer hydration state, CEC fluctuations, optimum IS configuration, crystallite size/distribution, average layer number per crystallite, and structural heterogeneities [29]. Despite the various works carried out that rely on the modeling of nanoclay structure, to our knowledge, no study has described the details of modeling and strategy execution. In some cases, the modeling method is briefly described without details [21, 25–26, 29, 37]. Hence there are always shortcomings to discover. This work focuses on a detailed description of the diffractogram modeling strategy of nanoclay structures based on the comparison of experimental reflections of 00l with those calculated, which makes it possible to reconstruct a theoretical model describing the layers stacking at the crystallite along the c*.

2. Materials and methods

2.1 Baseline material

A standard dioctahedral smectite SWy-2 extracted from the cretaceous formations of Wyoming (USA) and provided by the clay mineral repository is selected for the present study [38]. The structural formula per half-cell is given by [39]:

$(Si^{4+}{}_{3.96}, Al^{3+}{}_{0.04})$ $(Al^{3+}{}_{1.53}, Fe^{3+}{}_{0.18}, Fe^{2+}{}_{0.045}, Mg^{2+}{}_{0.26}, Ti^{4+}{}_{0.01})$ O_{10} $(OH)_2$ $(Ca^{2+}{}_{0.07}, K^{+}{}_{0.01}$ $Na^{+}{}_{0.2})$

This bentonite exhibits a low octahedral charge and extremely limited tetrahedral substitutions. The clay cation exchange capacity (CEC) is 101 meq/100 g [39]. Pretreatment of the starting material consists of preparing Na-rich montmorillonite suspension (SWy-Na), is realized following a classic protocol detailed by [25]. A cation exchange process is carried out for Barium cation (Ba^{2+}) in order to obtain the second reference sample SWy-Ba. The experimental protocol established consists of applying a mechanical shake for 48 h, followed by centrifugation at 4000 rpm. This step is repeated five times to ensure process achievement. After recovery of the solid fraction, a series of washes with distilled water will take place to remove excess salt from chloride ions. To collect XRD diffraction data, oriented samples were prepared by placing the obtained suspensions on a glass slide at air dry for 24 h.

2.2 Qualitative analysis

Qualitative XRD analysis is an essential task for the identification of a given diffractogram. It allows us to simplify the modeling approach. The following parameters are determined:

- The positioning of the peaks (2θ°).
- The basal spacing distance of the first reflection 001, to identify the hydration state (0 W, 1 W, 2 W...) of the sample using Bragg's law.
- FHWM values and rationality ξ [40].
- The crystallite size D_{hkl} in the perpendicular direction to the lattice planes [41] knowing that [8] shows that the phyllosilicates with multiscale structures begin from the nanoclay fraction to a micro agglomerate scale (> 10 μm; **Figure 1**). In addition, the reflection shape is checked with the aim of clearing the existence of a heterogeneous structure (possible reflection decomposition on two or more hydration states).

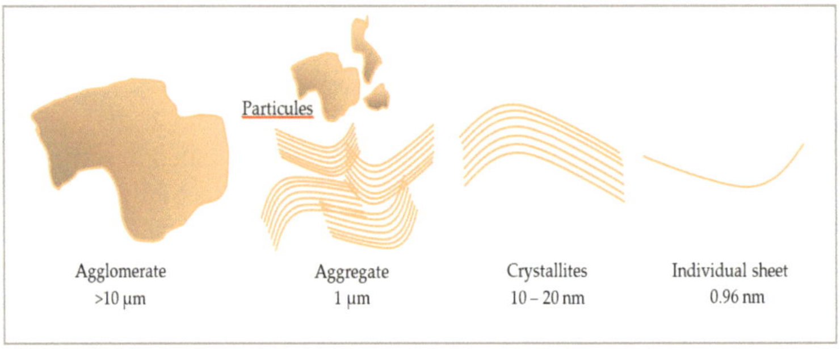

Figure 1.
Scheme of a clay particle at different scales.

2.3 Mathematical formalism: modeling (001) reflections

Theoretical diffracted intensity of lamellar structures generally based on powder diagrams is reported in **Figure 2**. Considering the case of a powder formed by crystallites having a layer with large lateral extension, the diffracted XRD intensity is expressed by:

$$I(s) = \frac{1}{S^2}\sum_{M_1}^{M_n} \alpha\,(M)\ \mathrm{Tr}\left\{\ \mathrm{Re}\left\{[F][W]\left([I] + 2\sum_{n}^{M-1}[Q]^n\right)\right\}\right\}, \tag{1}$$

where **Tr** is the trace of the matrix; **Re** is the real part of the expression; **M** is the number of layers per stack; **α(M)** is the weight distribution in thicknesses of the layer, M varies from M_1 to M_2.

$$\vec{S} = \mathbf{h}\,\overrightarrow{\mathbf{a}^*} + \mathbf{k}\,\overrightarrow{\mathbf{b}^*} + \mathbf{l}\,\overrightarrow{\mathbf{c}^*} \tag{2}$$

is the vector of the module reciprocal space

$$S = (2\,\mathbf{sin}\ \mathbf{sin}\,\boldsymbol{\theta})/\boldsymbol{\lambda};\ [\mathbf{1}] \tag{3}$$

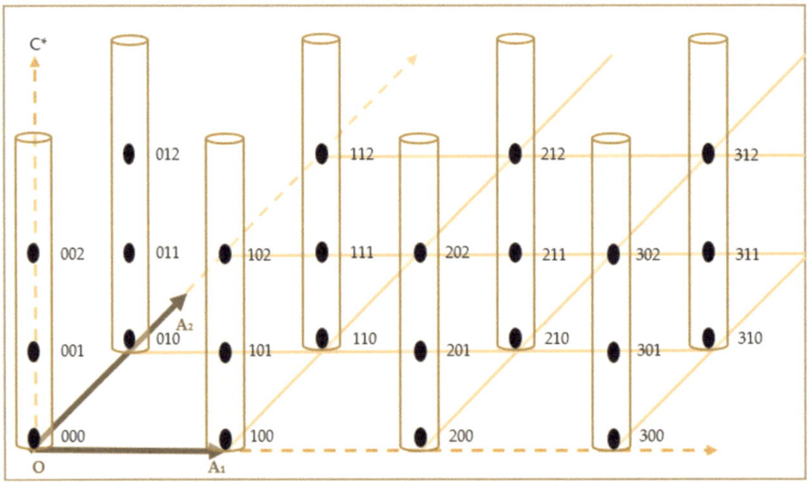

Figure 2.
Diffracted Intensity distribution in the case of layered structure.

is the identity matrix.

Matrix [F]: Matrix of structural factors, each element (ij) corresponds to the passage from a layer of type *i* to a neighboring layer *j*, this is expressed as

$$F_{ij}F_i(z).F_j^*(z). \tag{4}$$

$$[F] = \begin{bmatrix} F_0F_0^* & F_0F_1^* & .. & F_0F_g^* \\ F_1F_0^* & F_1F_1^* & .. & F_1F_g^* \\ . & . & .. & . \\ . & . & .. & . \\ F_gF_0^* & F_gF_1^* & .. & F_gF_g^* \end{bmatrix}. \tag{5}$$

For a given pair of Miller hk indices, the structure factor is calculated using the following expression:

$$F_{hk}(z) = \sum_n f_n e^{-2i\pi\left(hx_n + ky_n + lz_n\right)} \tag{6}$$

While the individual structural factors F_1, F_2 ... are obtained from the same relation using the same 2D network rotated from angle 2п/v, 4п/v, ... compared wih the fixed repository. Also, x_n and y_n are the coordinates of the atoms along the layer and z_n is their position in Å.

Matrix [W]: It is a diagonal matrix of order g, avec

$$\sum_{i=1}^{g} W_i = 1 \tag{7}$$

(W_i layer proportions of each stacking type).

$$[W] = \begin{bmatrix} W_1 & 0 & . & . & 0 \\ 0 & W_2 & . & . & 0 \\ 0 & 0 & . & . & 0 \\ 0 & 0 & . & . & 0 \\ 0 & 0 & . & . & W_g \end{bmatrix}. \tag{8}$$

Matrix $[Q]^n$: It characterizes the interference between diffracted waves by adjacent layers.

$$[Q]^n = \begin{bmatrix} P_{11}\varphi_{11} & P_{12}\varphi_{12} & . & . & P_{1g}\varphi_{1g} \\ P_{21}\varphi_{21} & P_{22}\varphi_{22} & . & . & P_{2g}\varphi_{2g} \\ . & . & . & . & . \\ . & . & . & . & . \\ P_{g1}\varphi_{g1} & P_{g2}\varphi_{g2} & . & . & P_{gg}\varphi_{gg} \end{bmatrix}, \tag{9}$$

where $\mathbf{P_{ij}}$ is the conditional probability that a type i sheet is followed by a type j sheet. $\boldsymbol{\varphi_{ij}}$ is the phase shift between two waves $\left\{ \varphi_{ij} = e^{2i\pi \vec{s}\,\vec{t_{ij}}} \right\}$. $\vec{\mathbf{t_{ij}}}$ is the translation between sheet i and sheet j first neighbor.

The relationships between probabilities and abundances are given by [7, 13, 37]:

$$\sum_{i=1}^{g} P_{ij} = 1 \tag{10}$$

$$\sum_{j=1}^{g} W_j P_{ji} = W_i \tag{11}$$

The theoretical XRD profile allows us to determine [37]: (i) the layers succession law within the stack; (ii) the number of different types of layers; (iii) relative layers type abundances; (iv) the average number of layers per stack; and (v) the weight distribution in thickness of the stacks.

The quasi-homogeneous model that assumes Markovian statistics is tremendously utilized [34, 37]. For illustration purposes, one may consider the case of mixed layers containing two types (A and B) and different sets of junction probability parameters. It is an interaction with the first neighbors since the translation between layers depends only on the nature of the previous layer or the next layer. In this type of interstratification, two main trends appear (**Table 1**).

Case of the random system that corresponds to aleatory layers succession within the stack:

$$\begin{cases} W_A = P_{AA} = P_{BA} \\ W_B = P_{BB} = P_{AB} \end{cases}. \tag{12}$$

For all cases, the matrices [Q] and [W] have the following forms:

$$[Q]^n = \begin{bmatrix} P_{AA}\varphi_{AA} & P_{AB}\varphi_{AB} \\ P_{BA}\varphi_{BA} & P_{BB}\varphi_{BB} \end{bmatrix} \tag{13}$$

$$[W] = \begin{bmatrix} W_A & 0 \\ 0 & W_B \end{bmatrix} \tag{14}$$

With

$$\varphi_{AB} = e^{2i\pi \vec{s}\,\overrightarrow{t_{AB}}} \tag{15}$$

$\overrightarrow{t_{AB}}$ is the translation between the first neighboring layers.

Relative abundances and junction conditional probabilities are linked by the following expressions:

$$W_A + W_B = 1 \tag{16}$$

$$P_{AA} + P_{AB} = 1 \tag{17}$$

$$P_{BA} + P_{BB} = 1, \tag{18}$$

$$\begin{cases} W_A = W_A P_{AA} + W_B P_{BA} \\ W_B = W_A P_{AB} + W_B P_{BB} \end{cases} \rightarrow W_A P_{AB} = W_B P_{BA}. \tag{19}$$

All possible types of layer stacks are characterized in (**Figure 3**).

In the case of a random stack, no stacking sequence is prohibited, so the probability of a layer appearing in a sequence depends only on its abundance. Thus $P_{AA} = W_A$ and this type of stacking is characterized in **Figure 3** by an increasing linear function. In the opposite case, where the succession of the two layers with different nature is prohibited

Tendency to segregation	Tendency to regularity
$\begin{cases} W_A < P_{AA} \leq 1 \\ W_B < P_{BB} \leq 1 \end{cases}$ (20) With total demixing for $P_{AA} = P_{BB} = 1$	$\begin{cases} W_A < P_{BA} \leq 1 \\ W_B < P_{AB} \leq 1 \end{cases}$ (21)

Table 1.
Segregation and regularity layers stacking tendency.

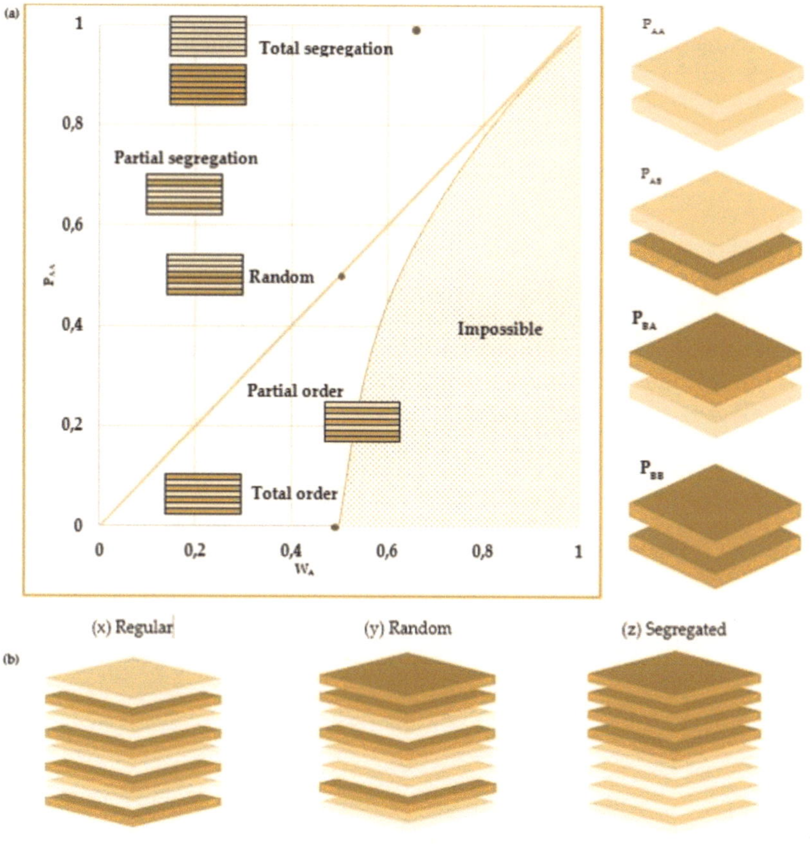

Figure 3.
(a) Junction probability diagram for two-component mixed layers (adapted from [42]). W_A and P_{AA} represent the relative proportion of A layers and the probability of finding an A layer after an A layer, respectively. Specific junction probabilities correspond to random interstratification. Possible stacking sequences corresponding to the different cases are schematized. Between the probability of occurrence of a layer B following a layer A as a function of the proportion of A layer [42]. (b) Type of stack deduced according to the parameters reported on the diagram: series (x) Regular, (y) random, and (z) segregated.

we have; $P_{AB} = P_{BA} = 0$. The probability that two layers of the same type will succeed each other is equal to the unit: $P_{AA} = P_{BB} = 1$. Then it is no longer a question of interstratification because the two types of layers no longer coexist at the heart of the same crystal but of total segregation or total separation. The maximum order is defined in the case of a prohibited succession between two minority layers. For example, if the type B layer is a minority, $P_{BB} = 0$, based on the two relationships below:

$$\begin{cases} \mathbf{P_{BA}} = \mathbf{1} \\ \mathbf{P_{AB}} = \dfrac{\mathbf{W_B}}{\mathbf{W_A}} \end{cases} \rightarrow \mathbf{P_{AA}} = \mathbf{1} - \frac{\mathbf{W_B}}{\mathbf{W_A}} = \frac{\mathbf{2W_A - 1}}{\mathbf{W_A}} \tag{22}$$

3. Implemented code

3.1 Code description

The modeling program is a "Fortran" code developed and improved by several authors [7, 8, 10] based on the mathematical formalism detailed by [37]. It makes it possible to create theoretical X-ray diffractograms in the case of lamellar structures. The quality control agreement between the two profiles is carried out separately using a calculated R_{WP} and/or Rp confidence factor [21, 29].

3.2 Code architecture

The program is essentially divided into three parts, respectively, (i) XRD parameters (experimental conditions and technical parameters related to the diffractometer), (ii) layered materials specification, and (iii) intrinsic structural properties (layers distribution, average number of layer per crystallite, and layer distribution function). The basic architecture of the executable file is reported in **Figure 4**.

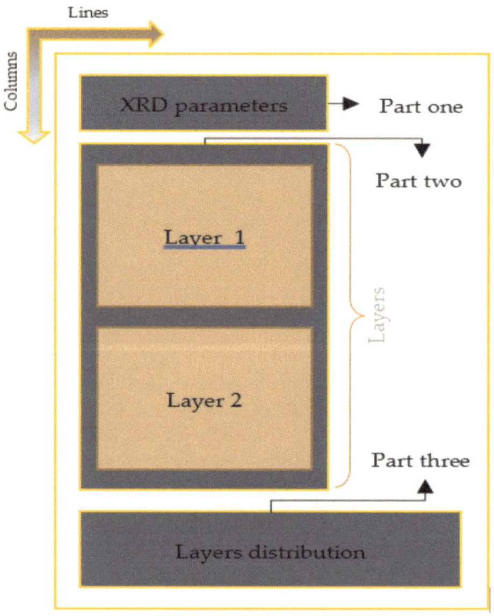

Figure 4.
Basic architecture of the executable file.

3.3 Input parameters affecting theoretical X-rays diffractograms: case of nanoclay

3.3.1 Hydration state

The nanoclay hydration state is simply defined by the amount of water inserted into the IS which induces an increase in the layer thickness [43]. A discrete simple hydration state is detailed in **Figure 5**.

The theoretical profile for each hydration state is reported on **Figure 6**. A logical translation toward low 2θ values is accompanied by an increase in the layer thickness input (according to Bragg Brentano's law). The proposed structure in this case is composed of two types of layers with a major contribution (100%) of the first layer, neglecting the second.

3.3.2 Water molecule abundance (nH_2O)

The water molecule amount variation has a direct impact on the increase and/or decrease in the relative diffracted intensity essentially for the higher diffraction

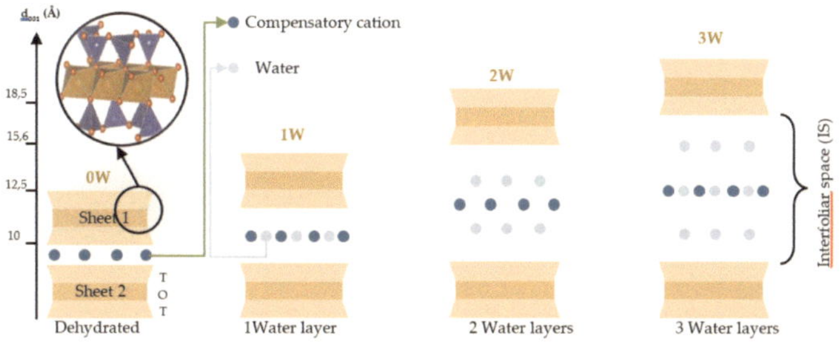

Figure 5.
Presentation of variables hydration states. The d_{001} basal spacing value is the projection of the period c on the normal to the layer.*

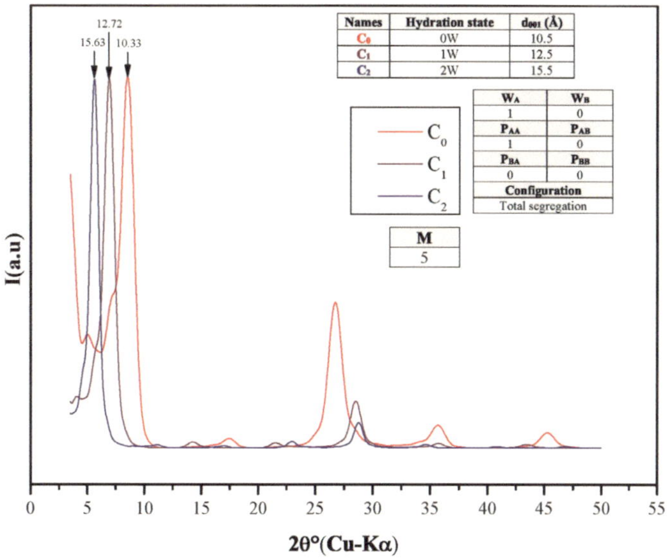

Figure 6.
Separated theoretical homogenous layer structures.

orders. An example of this impact is reported in **Figure 7**. A homogeneous 1 W hydration state is considered and a discrete water molecule distribution (0.1 increasing step) is directed.

A remarkable reflection intensity evolution is visualized only by increasing the amount of water molecules in the IS from 0.1 to 1 (**Figure 8**) while maintaining the same reflection position.

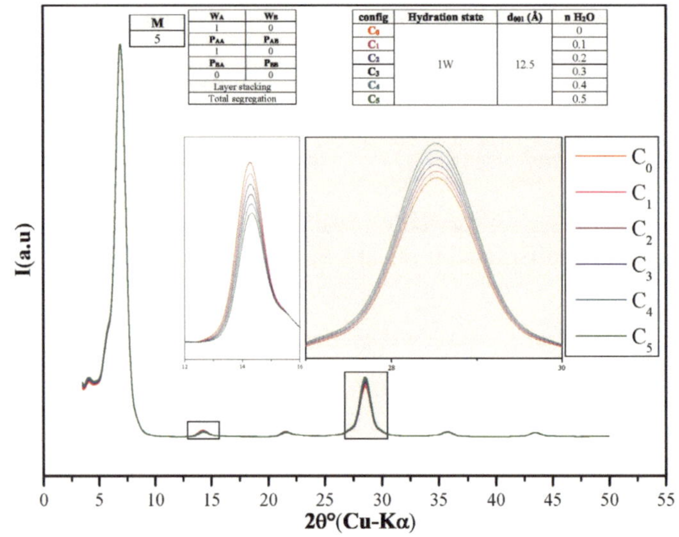

Figure 7.
Effect of varying water molecule abundance on the theoretical XRD profiles.

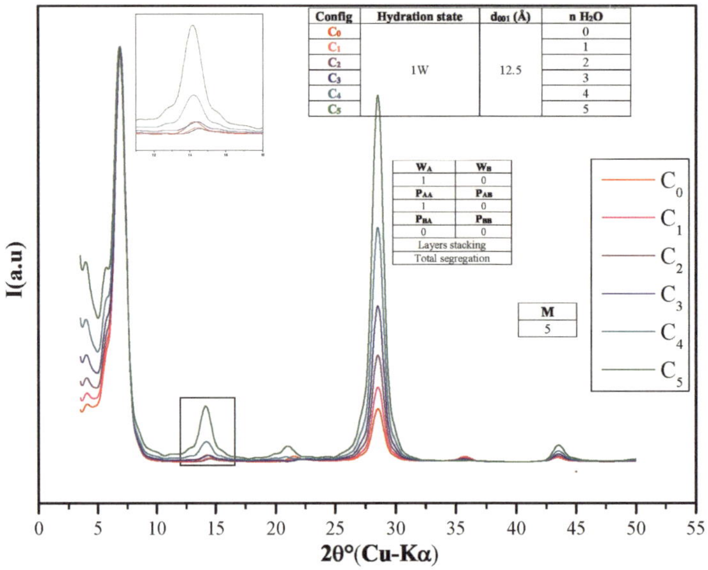

Figure 8.
Relative reflection intensity evolution by increasing IS water molecule abundance.

3.3.3 Variation of the average number of layers M per stack

The variation of the average number of layers M per stack affects the theoretical profiles (**Figure 9**), essentially the FWHM increases dramatically when the average number of layers is reduced. Note that when the average number of layers M decreases, the width of the lines 00ℓ (with $\ell = 1, 2, \ldots$ N) increases and vice versa.

3.3.4 Layer abundance and the probabilities junction law (W_i and P_{ij})

Variations in the relative abundances W_i and P_{ij} probabilities of layers stacking make it possible to visualize essentially three types of stacks: (i) random, (ii) regular, and (iii) segregated. Two-layer types were considered (i.e. 0 W (anhydrate) and 1 W (hydrated).

A **random distribution** refers to the chaotic stacking mode between two different types of layers (**Figure 10**). Note that the simultaneous increase of the W_A and P_{AA} values (with $W_A = P_{AA}$) provoke a 001 reflection shift while keeping the same relative intensity. For the second reflection 002 (~14° (2θ)), we also visualize displacement toward the low angles accompanied by a slight increase in diffracted intensity. The intensity of the 4th reflection order (position at 28° (2θ)) present a fluctuation versus W_A and P_{AA} value.

The **regular distribution** presents a well-ordered stacking mode between two different layers types (**Figures 11** and **12**). The theoretical reflections of this distribution exhibit the same behavior for random configuration by increasing the values of W_A and P_{AA}.

The **segregated distribution** involved an intermediate state between the regular and the random distribution. It is a mixture at short range between regular and chaotic layer stacking mode. The graphical effect of this distribution type on the theoretical diffracted intensity is reported in **Figures 13** and **14**.

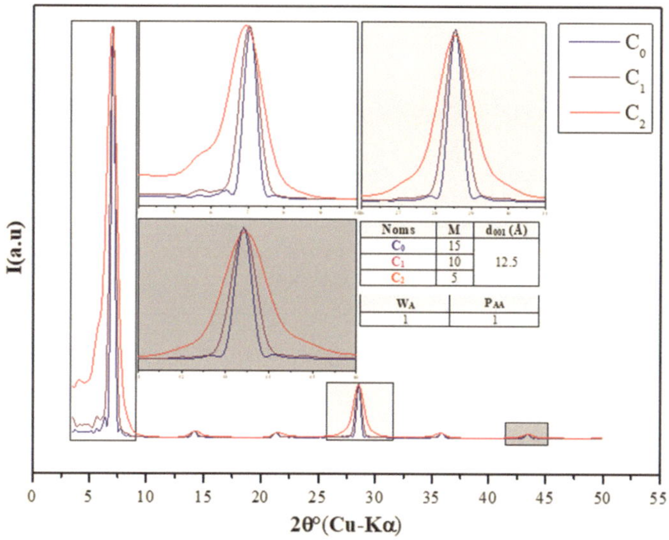

Figure 9.
Effect of the variation of the average number of layers M per stack on the theoretical profiles shape in the case of total segeregation.

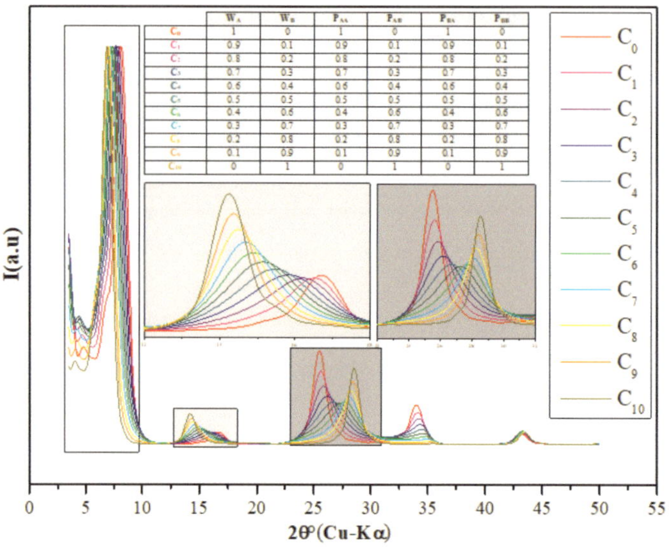

Figure 10.
Theoretical profiles of the variation of relative weights and probabilities (random configuration).

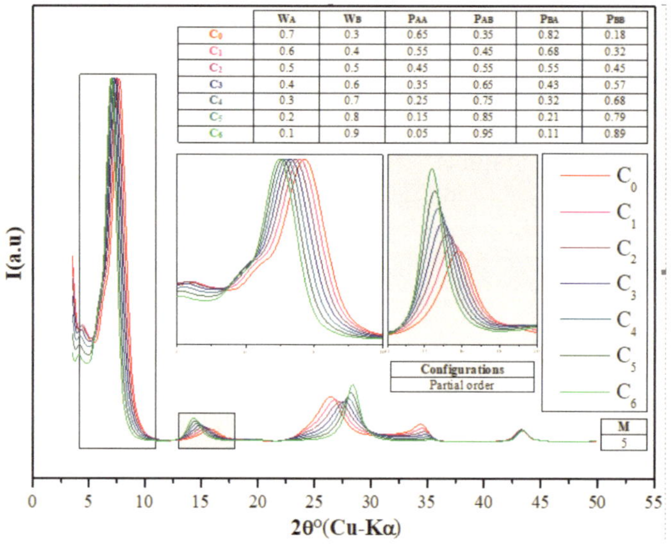

Figure 11.
Theoretical profiles of the change in relative abundance and junction probabilities law in the case of partial order stacking trend.

4. Modeling strategy

Before starting the modeling, several technical parameters such as improvement of the diffraction experimental data acquisition, chemical formula (extracted from literature or other analytical technics), experimental diffractions conditions, layer composition, atomic and onic scattering factors [44], and atoms coordinates must be checked and controlled in order to minimize the input variables thereafter. The

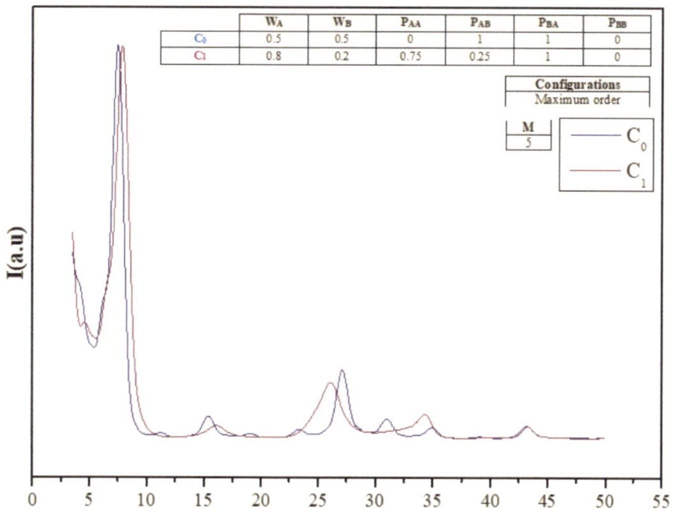

Figure 12.
Theoretical profiles of the change in relative abundance and junction probabilities law in the case of maximum order stacking.

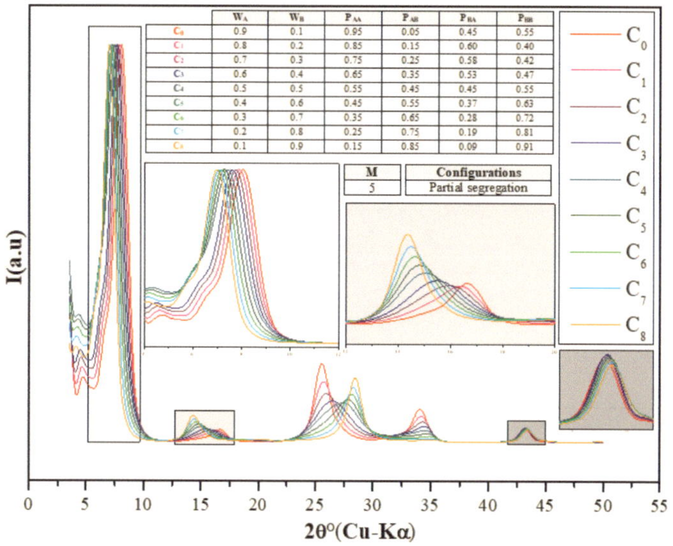

Figure 13.
Theoretical profiles change when varying relative abundance and junction probabilities law in the case of partial segregation trend.

fitting strategy consists of reproducing the experimental XRD pattern using a main homogeneous structure.

If necessary, additional contributions to the diffracted intensity are introduced to account for improve agreement between calculated and experimental patterns (i.e., if we have more one main structure, a MLS can be introduced). Indeed, the main 001 reflection can be decomposed into several theoretical weighted phases (**Figure 15**). The presence of two MLSs does not imply that two populations of particles are physically present in the sample [21, 45, 46]. Therefore, layers with the same hydration state present in the different MLSs contributing to the diffracted intensity are

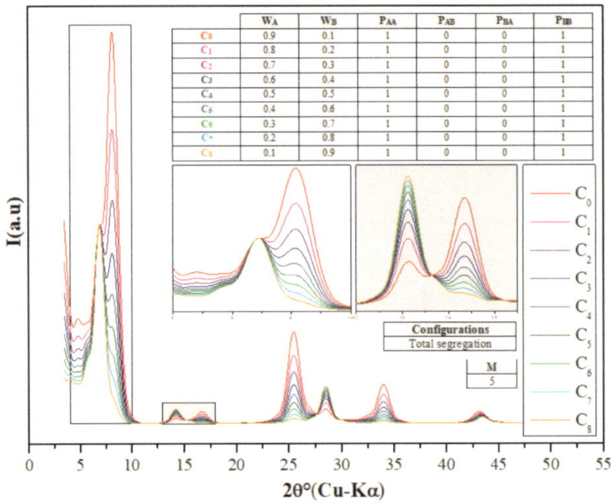

Figure 14.
Theoretical profiles change when varying relative abundance and junction probabilities law in the case of total segregation.

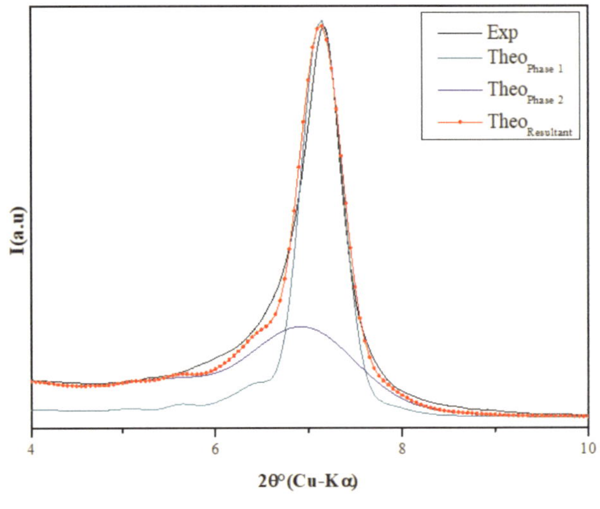

Figure 15.
Theoretical decomposition of the main 001 reflection.

assumed to have identical properties (chemical composition, layer thickness, and z coordinates of atoms).

Agreement between theoretical and experimental XRD profile is evaluated by the calculation of the R_{WP} trust factor based on the expression quoted [47] which must be around 5%.

$$\mathbf{R_{WP}} = \sqrt{\frac{\sum\left(\mathbf{I}(2\theta_i)_{exp} - \mathbf{I}(2\theta_i)_{theo}\right)^2}{\sum\left(\mathbf{I}(2\theta_i)_{exp}\right)^2}} \times \mathbf{100\%}. \quad (23)$$

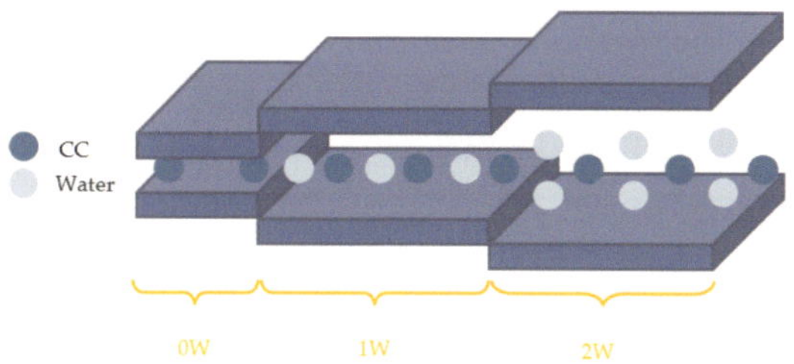

Figure 16.
Example of intra-layer hydration heterogeneity.

Similarly, there is another alternative that involves combining two or more hydration states in the same layer (interstratification intra-layer) by varying the percentage of their abundance (**Figure 16**). This physically results in fluctuations in the layer thickness as a function of the (0kl) surface which affects the properties of the clay particle for lateral extension (a and b $\sim \infty$). Although this method is easier than theoretical decomposition, it is not recommended because the width of the lateral extension cannot be determined qualitatively.

5. Modeling example

This paragraph is intended to mention the modeling details for two different samples SWy-Na and SWy-Ba, where the CEC is fully saturated by Na^+ and Ba^{2+} cations.

5.1 SWy-Na sample

Qualitative XRD analysis shows a homogeneous 1 W hydration state (12.33 Å; **Figure 17**). A low FWHM and ξ parameter value confirms the symmetric shape of the 001 reflection (**Table 2**).

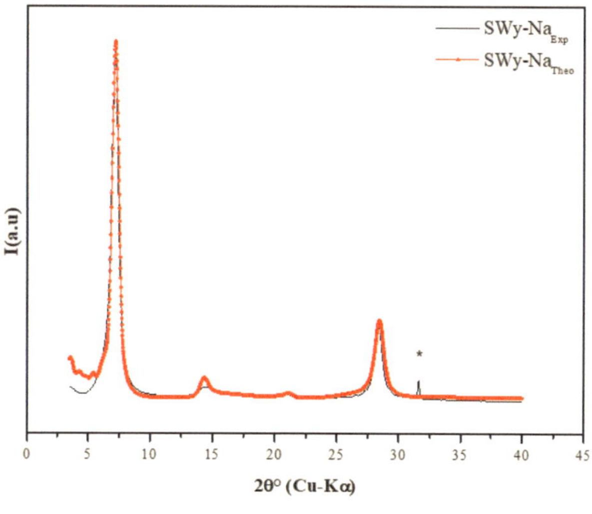

Figure 17.
*Best agreement between theoretical and experimental XRD profiles obtained in the case of SWy-Na. * Halite (NaCl). The sample used here is Wyoming montmorillonite (SWy-2) exchanged with sodium (SWy-Na) during the purification procedure. All the details dismaying this sample are mentioned in [26].*

Sample	2θ°	d_{001} (Å)	FWHM (2θ°)	D (Å)	ξ (Å)	Character
SWy-Na	7.16	12.33	0.74	18.77	0.062	Homogeneous
SWy-Ba	6.83	12.93	1.15	12.08	0.397	Heterogeneous

Notes: 2θ°, Bragg's angle; d_{001}, basal distance of the 1st reflection; D, average crystalline size; FWHM, full-width at half maximum; ξ, deviation from rationality.

Table 2.
Qualitative XRD investigation in the case of the SWy-Na and SWy-Ba sample.

Sample	d_{001}	nH_2O	W_A	P_{AA}	S_M	M	C	R_{WP} (%)
SWy-Na	10.5	0	0.80	0.85	$R_{1\text{-seg}}$	12	He	2.17
	12.5	2						
SWy-Ba	12.5	1	0.8	0.9	$R_{1\text{-seg}}$	8	He	1.34
	15.5	1.5						
		1.5						

Notes: *d_{001}, interlamellar distance; nH_2O, number of water molecules per half-cell; zH_2O, position of the molecules along the c axis of the H_2O molecule is attached to 9.6 Å for hydration states 1 W; the position of the exchangeable cations per half-cell calculated along the axis c^* is fixed à 9.6 Å for hydration states 1 W; M, average number of sheets per stack; S_M, layer stacking mode; R_0, maximum order; R_1, random stacking; $R_{1\text{-}X}$, associated stacking (segregation-partial order); C, characters; Ho, homogeneous; He, heterogeneous; R_{WP}, confidence factor.*

Table 3.
Structural parameters extracted from XRD modeling approach.

The MLS used to achieve the best agreement between the calculated and experimental XRD model (**Figure 17**), shows a heterogeneous hydration character that results in the coexistence of two layers types (0 W and 1 W) with a strong dominance for the 1 W phases about 80%. This contradicts the qualitative analysis which indicates that the structure is purely homogeneous (a pure homogeneous 1 W hydration state). The structural parameters are summarized in **Table 3**. The confidence factor R_{WP} is very low (2.17%), which reflects the high fit quality.

5.2 SWy-Ba sample

The SWy-Ba sample present a two-water hydration state 2 W [26, 46, 48]. A heterogeneous hydration behavior is the main description based on the asymmetric XRD reflection profile shape and the elevated FWHM/ξ parameter value (**Table 2**, **Figure 18**). The optimized theoretical model (**Figure 18**) is obtained by combining MLS with variable hydration state. The coexistence of two types of sheets (1 W and 2 W) with a major 1 W layer fraction. The existence of two types of compensating cations is probably related to the incompletion of the cation exchange process (sodium residue/excess salt). The structural parameters are summarized in **Table 3**. The confidence factor R_{WP} is very low (1.34%), which reflects the high fit quality.

6. Conclusion

This work investigates the strength of the modeling X-ray diffractograms method used to rebuild theoretically the full structure along c* axis for the layered materials such as nanoclays. This technic is an indirect method based on the comparison

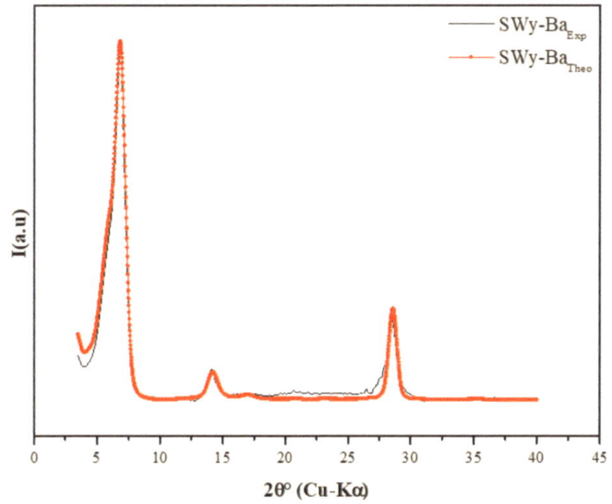

Figure 18.
Best agreement between theoretical and experimental XRD profiles obtained in the case of SWy-Ba.

between experimental and theoretical profiles. An intrinsic mathematical formalism was detailed in this chapter. The implemented code used to perform this operation has been described in detail by explaining the role of each input parameters such as:

i. The abundance and the junction probability law describing the layer stacking mode (W_A and P_{AA}).

ii. The effect of fluctuations of the number of layers per crystallite M in shape and the response of the theoretical diffracted intensity.

iii. The water molecule nH_2O amount effect in the improvement of the reflection intensity agreement for the higher diffraction orders (>001).

iv. Role of R_{WP} confidence factor to validate the obtained model (≈5%).

A complete modeling strategy is detailed and accompanied by an application on two nanoclay specimen saturated, respectively, by Na+ and Ba2+ cations.

Conflict of interest

The authors declare no conflict of interest.

Author details

Walid Oueslati*, Chadha Mejri and Abdesslem Ben Haj Amara
Université de Carthage, Faculté des Sciences de Bizerte, LR19ES20: Ressources, Matériaux, Et Ecosystémes (RME), Bizerte, Tunisia

*Address all correspondence to: walidoueslati@ymail.com

References

[1] Kittrick JA. Soil minerals in the Al2O3-SiO2-H2O system and a theory of their formation. Clays and Clay Minerals. 1969;**17**(3):157-167. DOI: 10.1346/CCMN.1969.0170304

[2] Kittrick JA. Interlayer forces in montmorillonite and vermiculite. Soil Science Society of America Journal. 1969;**33**(2):217-222. DOI: 10.2136/sssaj1969.03615995003300020017x

[3] Laird DA. Model for crystalline swelling of 2: 1 phyllosilicates. Clays and Clay Minerals. 1996;**44**(4):553-559. DOI: 10.1346/CCMN.1996.0440415

[4] Laird DA. Layer charge influences on the hydration of expandable 2: 1 phyllosilicates. Clays and Clay Minerals. 1999;**47**(5):630-636. DOI: 10.1346/CCMN.1999.0470509

[5] Glaeser R, Mantine I, Méring J. Observations sur la beidellite. Bulletin du Groupe Français des Argiles. 1967;**19**(1):125-130. DOI: 10.3406/argil.1967.1069

[6] Ben Brahim J, Besson G, Tchoubar C. Etude des profils des bandes de diffraction X d'une beidellite-Na hydratée à deux couches d'eau. Détermination du mode d'empilement des feuillets et des sites occupés par l'eau. Journal of Applied Crystallography. 1984;**17**(3):179-188. DOI: 10.1107/S0021889884011262

[7] Plançon A. Diffraction by layer structures containing different kinds of layers and stacking faults. Journal of Applied Crystallography. 1981;**14**(5):300-304. DOI: 10.1107/S0021889881009424

[8] Rhaiem HB, Pons CH. Obtention de la fonction de modulation experimentale pour les gels de smectites contenant du fer. Clay Minerals. 1984;**19**(2):257-260. DOI: 10.1180/claymin.1984.019.2.14

[9] Sato T, Watanabe T, Otsuka R. Effects of layer charge, charge location, and energy change on expansion properties of dioctahedral smectites. Clays and Clay Minerals. 1992;**40**(1):103-113. DOI: 10.1346/CCMN.1992.0400111

[10] Ben Haj Amara A, Ben Brahim J, Plançon A, Ben Rhaiem H. Etude par diffraction X des Modes d'Empilement de la Nacrite Hydratée et Deshydratée. Journal of Applied Crystallography. 1998;**31**(5):654-662. DOI: 10.1107/S0021889898000363

[11] Drits VA, Plançon A, Sakharov BA, Besson G, Tsipursky SI, Tchoubar C. Diffraction effects calculated for structural models of K-saturated montmorillonite containing different types of defects. Clay Minerals. 1984;**19**(4):541-561. DOI: 10.1180/claymin.1984.019.4.03

[12] De la Calle C, Plançon A, Pons CH, Dubernat J, Suquet H, Pezerat H. Mode d'empilement des feuillets dans la vermiculite sodique hydratee a une couche (phase a 11· 85 Å). Clay Minerals. 1984;**19**(4):563-578. DOI: 10.1180/claymin.1984.019.4.04

[13] Plançon A. Order-disorder in clay mineral structures. Clay Minerals. 2001;**36**(1):1-14. DOI: 10.1180/000985501547286

[14] Plançon A. New modeling of X-ray diffraction by disordered lamellar structures, such as phyllosilicates. American Mineralogist. 2002;**87**(11-12):1672-1677. DOI: 10.2138/am-2002-11-1216

[15] Sato H, Suzuki S. Fundamental study on the effect of an orientation of clay particles on diffusion pathway in compacted bentonite. Applied Clay Science. 2003;**23**(1-4):51-60. DOI: 10.1016/S0169-1317(03)00086-3

[16] Iwasaki T, Watanabe T. Distribution of Ca and Na ions in dioctahedral smectites and interstratified dioctahedral mica/smectites. Clays and Clay Minerals. 1988;**36**(1):73-82. DOI: 10.1346/CCMN.1988.0360110

[17] Glaeser R, Méring J. lsothermes d'hydratation des montmorillonites bi-ioniques (Na, Ca). Clay Minerals Bulletin. 1954;**2**(12):188-193. DOI: 10.1180/claymin.1954.002.12.13

[18] Bérend I, Cases JM, François M, Uriot JP, Michot L, Masion A, et al. Mechanism of adsorption and desorption of water vapor by homoionic montmorillonites: 2. The Li+ Na+, K+, Rb+ and Cs+—exchanged forms. Clays and Clay Minerals. 1995;**43**(3):324-336. DOI: 10.1346/CCMN.1995.0430307

[19] Calarge L, Lanson B, Meunier A, Formoso ML. The smectitic minerals in a bentonite deposit from Melo (Uruguay). Clay Minerals. 2003;**38**(1):25-34. DOI: 10.1180/0009855033810075

[20] Claret F, Sakharov BA, Drits VA, Velde B, Meunier A, Griffault L, et al. Clay minerals in the Meuse-Haute Marne underground laboratory (France): Possible influence of organic matter on clay mineral evolution. Clays and Clay Minerals. 2004;**52**(5):515-532. DOI: 10.1346/CCMN.2004.0520501

[21] Ferrage E, Lanson B, Sakharov BA, Drits VA. Investigation of smectite hydration properties by modeling experimental X-ray diffraction patterns: Part I Montmorillonite hydration properties. American Mineralogist. 2005; **90**(8-9):1358-1374. DOI: 10.2138/am.2005.1776

[22] Ferrage E, Lanson B, Sakharov BA, Geoffroy N, Jacquot E, Drits VA. Investigation of dioctahedral smectite hydration properties by modeling of X-ray diffraction profiles: Influence of layer charge and charge location. American Mineralogist. 2007;**92**(10): 1731-1743. DOI: 10.2138/am.2007.2273

[23] Oueslati W, Rhaiem HB, Lanson B, Amara ABH. Selectivity of Na–montmorillonite in relation with the concentration of bivalent cation (Cu2+, Ca2+, Ni2+) by quantitative analysis of XRD patterns. Applied Clay Science. 2009;**43**(2):224-227. DOI: 10.1016/j.clay.2008.09.010

[24] Ammar M, Oueslati W, Rhaiem HB, Amara ABH. XRD profile modeling approach tools to investigate the effect of charge location on hydration behavior in the case of metal exchanged smectite. Powder Diffraction. 2013;**28**(S2):S284-S300. DOI: 10.1017/S0885715613000973

[25] Oueslati W, Meftah M. Discretization of the water uptake process of Na-montmorillonite undergoing atmospheric stress: XRD modeling approach. Advances in Materials Science and Engineering. 2018; **2018**:5219624. DOI: 10.1155/2018/5219624

[26] Mejri C, Oueslati W, Amara ABH. How the solid/liquid ratio affects the cation exchange process and porosity in the case of dioctahedral smectite: Structural analysis. Adsorption Science & Technology. 2021;**2021**:9732092. DOI: 10.1155/2021/9732092

[27] Drits VA, Lindgreen H, Salyn AL. Determination of the content and distribution of fixed ammonium in illite-smectite by X-ray diffraction:

Application to North Sea illite-smectite. American Mineralogist. 1997;**82**(1-2): 79-87. DOI: 10.2138/am-1997-1-210

[28] Drits VA. Structural and chemical heterogeneity of layer silicates and clay minerals. Clay Minerals. 2003;**38**(4):403-432. DOI: 10.1180/0009855033840106

[29] Lanson B. Modelling of X-ray diffraction profiles: Investigation of defective lamellar structure crystal chemistry. EMU Notes in Mineralogy. 2011;**11**(4):151-202. DOI: 10.1180/EMU-notes.11

[30] Hendricks S, Teller E. X-ray interference in partially ordered layer lattices. The Journal of Chemical Physics. 1942;**10**(3):147-167. DOI: 10.1063/1.1723678

[31] Méring J. L'interférence des rayons X dans les systèmes à stratification désordonée. Acta Crystallographica. 1949;**2**(6):371-377. DOI: 10.1107/S0365110X49000977

[32] Kakinoki J, Komura Y. Diffraction by a one-dimensionally disordered crystal I. The intensity equation. Acta Crystallographica. 1965;**19**(1):137-147. DOI: 10.1107/S0365110X65002888

[33] Cesari M, Allegra G. The intensity of X-rays diffracted by monodimensionally disordered structures. Case of identical layers and three different translation vectors. Acta Crystallographica. 1967; **23**(2):200-205. DOI: 10.1107/S0365110X67002464

[34] Drits VA, Sakharov BA. X-ray structural analysis of mixed-layer minerals. Transactions of the USSR Academy of Sciences. 1976;**295**:1-252

[35] Sakharov BA, Naumov AS, Drits VA. X-ray diffraction by mixed-layer structures with a random distribution of stacking faults. Soviet Physics Doklady. 1982;**27**:523

[36] Watanabe T, Sato T. Expansion characteristics of montmorillonite and saponite under various relative humidity conditions. Clay Science. 1988;**7**(3): 129-138. DOI: 10.11362/jcssjclayscience1960.7.129

[37] Tsybulya SV, Cherepanova SV, Kryukova GN. Full profile analysis of X-ray diffraction patterns for investigation of nanocrystalline systems. In: Diffraction Analysis of the Microstructure of Materials. Berlin, Heidelberg: Springer; 2004. pp. 93-123. DOI: 10.1007/978-3-662-06723-9_4

[38] Moll WF Jr. Baseline studies of the clay minerals society source clays: geological origin. Clays and Clay Minerals. 2001;**49**(5):374-380

[39] Mermut AR, Cano AF. Baseline studies of the clay minerals society source clays: chemical analyses of major elements. Clays and Clay Minerals. 2001; **49**(5):381-386

[40] Bailey SW. Nomenclature for regular interstratifications. Clay Minerals. 1982;**17**(2):243-248. DOI: 10.1180/claymin.1982.017.2.09

[41] Holzwarth U, Gibson N. The Scherrer equation versus the 'Debye-Scherrer equation'. Nature Nanotechnology. 2011;**6**(9):534-534. DOI: 10.1038/nnano.2011.145

[42] Bethke CM, Altaner SP. Layer-by-layer mechanism of smectite illitization and application to a new rate law. Clays and Clay Minerals. 1986;**34**(2):136-145. DOI: 10.1346/CCMN.1986.0340204

[43] Dazas B, Ferrage E, Delville A, Lanson B. Interlayer structure model of

tri-hydrated low-charge smectite by X-ray diffraction and Monte Carlo modeling in the Grand Canonical ensemble. American Mineralogist. 2014; **99**(8-9):1724-1735. DOI: 10.2138/am.2014.4846

[44] Brown PJ, Fox AG, Maslen EN, O'Keefe MA, Willis BTM. Intensity of diffracted intensities. In: Fuess H, Hahn T, Wondratschek H, Müller U, Shmueli U, Prince E, et al., editors. International Tables for Crystallography. 2006. DOI: 10.1107/97809553602060000600

[45] Oueslati W, Rhaiem HB, Amara ABH. Effect of relative humidity constraint on the metal exchanged montmorillonite performance: An XRD profile modeling approach. Applied Surface Science. 2012;**261**:396-404. DOI: 10.1016/j.apsusc.2012.08.022

[46] Ammar M, Oueslati W, Rhaiem HB, Amara ABH. Effect of the hydration sequence orientation on the structural properties of Hg exchanged montmorillonite: quantitative XRD analysis. Journal of Environmental Chemical Engineering. 2014;**2**(3): 1604-1611. DOI: 10.1016/j.jece.2014.05.022

[47] Howard SA, Preston KD. Profile fitting of powder diffraction patterns. In: Bish DL, Post JE, editors. Modern Powder Diffraction. Berlin, Boston: De Gruyter; 2018. pp. 217-276. DOI: 10.1515/9781501509018-011

[48] Oueslati W, Ammar M, Chorfi N. Quantitative XRD analysis of the structural changes of Ba-exchanged montmorillonite: effect of an in situ hydrous perturbation. Minerals. 2015; **5**(3):507-526. DOI: 10.3390/min5030507

Chapter 2

New Trends in Clay-Based Nanohybrid Applications: Essential Oil Encapsulation Strategies to Improve Their Biological Activity

Houda Saad, Ameni Ayed, Mondher Srasra, Sameh Attia, Ezzeddine Srasra, Fatima Charrier-El Bouhtoury and Olfa Tabbene

Abstract

Essential oils (EOs) are used in medicinal, pharmaceutical, cosmetic, agricultural, and food industries thanks to their key properties and multiple benefits. Several techniques and embedding materials are used to nanoencapsulate EOs, in order to keep them from environmental conditions and boost their bioefficiency by controlled release. In recent years, the interest for clay nanoparticles as nanoencapsulation materials for EOs is increasing owing to their abundance in nature, low cost, inertness, and special structure. Thus, this chapter focuses on highlighting data and contributions dealing with EOs incorporation into nanoclay particles, their current applications and nanohybrid formation benefits on the stability, bioavailability, and sustained release of EOs. An overview about nanoclays used for EOs nanoencapsulation is highlighted in the beginning of this chapter followed by a brief description of EOs' chemical composition and properties.

Keywords: essential oils, clay, nanohybrids, encapsulation, biological activities, controlled release, pharmaceutics and cosmetic industries, packaging and coatings, bio-agrochemicals

1. Introduction

Essential oils (EOs) are secondary metabolites of diverse aromatic plants biosynthesized in different plant organs [1] that can be extracted from leaves, flowers, and fruits by hydrodistillation, solvent-solvent extraction, and liquid CO_2 extraction [2, 3]. The EOs' chemical composition is too complex. It is a mixture of natural volatile compounds, such as terpenes, phenols, ketones, aldehydes, alcohols, carotenoids, flavonoids, esters, and phenylpropanoid [4]. Thank to these variable bioactive molecules, EOs find uses as gastronomic, nutritional, organoleptic, antiulcer, antiaging, anticancer, antidepressant, antitussive, antipyretic, analgesic, larvicidal, insecticidal, etc. [5, 6]. Owing to their versatile biological properties and the increasing demand, by the consumers, for biobased products, EOs are amply used in pharmaceutics industry, cosmetics, food industry, food packaging, nutraceuticals, and even as agrochemicals [4, 7]. Nonetheless, their practice is constantly facing several barriers comprising the high volatility and high risk of degradation upon direct exposure to heat, humidity, light, and oxygen, intense odor and taste, dose-dependent toxicity, and hydrophobicity [1, 4, 8]. The nanoencapsulation technology has been recommended as an innovative approach to overcome the limitations of the EOs use, by enhancing their bioavailability and bioefficiency and protecting them from extreme conditions [9]. Currently, liposomes, polymeric nanoparticles, metal nanoparticles, and carbon nanotubes are some of the broadly used nanomaterials. Yet, they present some impediments of use, as they are sometimes extremely toxic and/or carcinogenic in nature even in low concentrations, very expensive to acquire, and need very complex preparation processes [10].

Clay nanoparticles represent a promising alternative to nanomaterials mentioned above. In soil science, the term "clay" is related to a material class with a particle size <2 μm in equivalent spherical diameter. "Nanoclays" are included in the clay fraction with particle size <100 nm in diameter. Soil nanoclays are commonly predominated by phyllosilicates and often include metal hydroxides and organic matter. Ultrasonication, ultracentrifugation, and energetic stirring may be used to isolate them from the clay fraction [11]. Their structure may be lamellar, fibrous, or tubular nature with a hydrophilic character. Since ancient time, clay minerals have been widely investigated by humans in many fields including medicine, pharmacy, ceramic, plastic, cracking catalyst industries, food and beverage, coatings industry, agrochemicals. They are also helpful in environmental protection and remediation. All this interest for nanoclay exploitation in various fields is mainly due to their abundance, low cost, ecofriendly nature, nontoxicity as well as their unique and specific structural physicochemical and thermal properties, including large surface area, surface electric charges, immense porosity, low density, inertness, thermal and chemical stability, ion exchange capacity, and high adsorption capacity [12–15].

In this context, this chapter is dedicated to the illustration and update of clay/EOs nanosystems application in diverse fields. Firstly, the main clays used for EOs encapsulation are highlighted. Then, EOs' chemical composition and properties are briefly described. Finally, data concerning clay/EOs nanosystems development and valorization in various applications are provided, while emphasizing the benefits associated with nanoencapsulation, namely bioactivity, stability against aggressive conditions, and controlled release.

2. Clay minerals

Clays are natural earth materials resulting from chemical weathering operations on the earth's crust [16]. These are minerals with very fine grain size smaller than two micrometers [17].

A clay deposit generally comprises of impurities, namely feldspar, quartz, volcanic dust, fossil fragments, heavy minerals, carbonates minerals, etc. [18].

2.1 Clay minerals structure

Clay minerals belong to phyllosilicates family and are constituted by a stacking of sheets. **Figure 1** illustrates the terminology used to define the layer silicates structure [19]. So, it can be established:

- A simple plane, which is constituted by atoms (like linked O or OH);
- A sheet results from a combination of planes (like a silica tetrahedral sheet);
- A layer results from a combination of sheets;
- The crystal is issued from the stacking of several layers.

Every layer is constituted of two fundamental units:

The tetrahedral sheet, designed as (T), whose basic building unit is a "silica tetrahedron." In each unit, one silicon atom is surrounded by four oxygens. The resulting silica tetrahedra units are connected horizontally by sharing oxygens anions, *via* covalent bonding, to produce a sheet of Si_2O_6 $(OH)_4$ structure.

The octahedral sheet, designed as (O), is built by the horizontal association of numerous octahedra, by covalent bonding. Each unit is composed of Al or Mg ions (and occasionally Fe ion) surrounded by six oxygen atoms or hydroxyl groups. This arrangement allows an eight-sided configuration labeled "octahedron." When the (O) is dominated by aluminum, two Al^{3+} cations are so needed to maintain the electrical neutrality and the sheet is so designated as "dioctahedral sheet." When magnesium predominates the (O), three Mg^{2+} cations are required, and the sheet is termed as "trioctahedral sheet" [15, 20–22].

The multiple associations between (T) and (O) and the numerous chemical substitutions induce different clay minerals with diverse physicochemical characteristics [16]. Consequently, three clay minerals categories are discerned:

- The 1:1clay minerals (or TO type) with the structural formula $[SiO_4Al_4O_{10}(OH)_8]$. Kaolinite, dickite, halloysite, amesite, and lizardite are included in this category. Each clay layer is formed by one (T) and one (O);
- The 2:1 clay minerals (or TOT type) are built by one (O) interposed between two (T). Smectite, bentonite, montmorillonite, palygorskite, sepiolite, glauconite, mica, vermiculite, and saponite are allied to this category [23, 24];

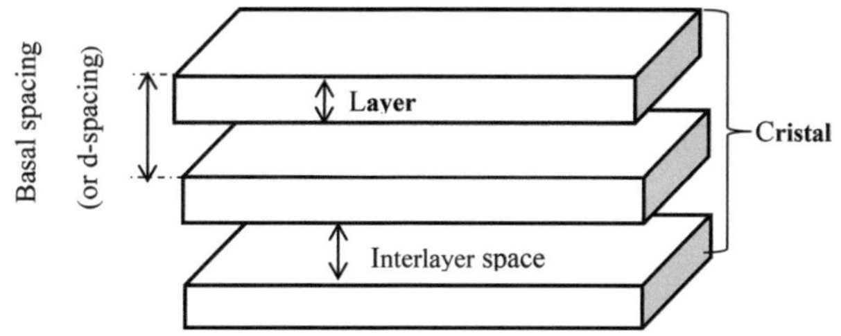

Figure 1.
A schematic illustration of the general structure of phyllosilicates.

- The space between two successive 1:1 and 2:1 layers is the "interlayer," which is empty if the layers are electrostatically neutral. If the layer takes an overabundance charge, defined as "layer charge," so it is neutralized by diverse interlayer elements, such as cations (Ca, Na, Mg, K), hydrated cations, and hydroxide octahedral groups. The hydroxide interlayer often forms an additional octahedral sheet yielding a 2:1:1 or TOTO layer. This structure is typical of chlorite, chamosite, and donbassite [25].

Clay minerals can be identified according to the assembly method and their shape. It is distinguished by the following:

- Fibrous clay minerals, such as palygorskite (attapulgite) and sepiolite;

- Tubular clay minerals, such as halloysite;

- Lamellar clay minerals, such as smectites and kaolinite [26].

2.2 Main clay minerals used for essential oil nanoencapsulation

2.2.1 Kaolin minerals

Kaolinite (Kaol) is the main element of this class. Its structure results from a combination of one tetrahedral sheet and one dioctahedral sheet. The basal spacing is about 7.1 Å [20]. It is differentiated by the presence of strong bonds between the layers, which opposes to its expansion and swelling [27]. Kaol is also characterized by negligible cation exchange capacity (CEC) and small surface area [28].

Halloysite clay minerals, with an empirical formula $Al_2Si_2O_5(OH)_{4n}H_2O$, belong to the Kaolin class, with the same chemical structure as kaolinite [29]. Based on crystalline and geographical conditions, multiple morphologies are defined. The most abundant one in nature is the tubular form resulting from the rolling of the kaolin sheets with the presence of a water molecules layer in the interlayer space [30, 31].

The interlayer water in halloysite nanotubes (HNTs) is one of the principal properties discerning HNTs from Kaol. The dehydration action causes irreversibly the change of the d_{001} spacing from 10 to 7 Å [32].

HNTs are composed of two kinds of hydroxyl groups according to their position. The inner hydroxyl groups are located between layers and the outer hydroxyl groups that are located on the surface of the HNTs. The most ones are inner groups due to the tubular form [33]. HNTs diameter is of 40–70 nm with inner lumen diameter of 10–15 nm and length of 1000–2000 nm [34].

HNTs are characterized by a relatively high specific surface area and a total pore volume that is much higher than that of platy Kaol. This should be attributed to the rich pores in the structure of halloysite [30].

2.2.2 Semectite minerals

Smectites are TOT phyllosilicates type. The association between smectite layers leads to a constant van der Waals break between the layers, known as "interlayer" [35].

Two smectite subgroups are existing:

i. Dioctahedral smectites, as montmorillonite, non-tronit, beidellite, etc.

ii. Trioctahedral smectites, as hectorite, saponite, sauconite, etc. [22].

The smectite clays structure is able to expand and depreciate without losing its crystallinity.

Smectite layers are typified by plentiful isomorphic substitutions in tetrahedral (Si^{4+} replaced by Al^{3+}) and octahedral (Al^{3+} replaced by Mg^{2+} or Fe^{2+} and Mg^{2+} substituted by Li^{+}) positions generating a layer charge that is neutralized by hydrated cations in the interlayer space, named interlayer cations (Na^{+}, Ca^{2+}, K^{+}, etc.) [36].

Montmorillonite (MMT) is the most important smectite clay owing to its commercial value. It emanates from the replacement of the Si^{4+} in the silica tetrahedral sheets by Al^{3+} and Al^{3+} in the alumina octahedral sheets by Mg^{2+} leading to a negative charge that is neutralized by the Na^{+} and Ca^{2+} interlayer cations.

These cations can be facilely supplied by other organic or inorganic cations, due to the incomparable hydrophilicity, swelling, adsorption, and fluidity properties of montmorillonite [37].

2.2.3 Palygorskite and sepiolite minerals

Sepiolite (Sep) [$Mg_8Si_{12}O_{30}(OH)_4(OH2)_{4.n}(R_2 + (H_2O)_8$] and **palygorskite** (Pal) [$MgAl_3Si_8O_{20}(OH)_3(H_2O)_{4.n}(R_2 + (H_2O)_4$] are 2:1 layer silicates being distinct from other clay minerals owing to the existence of continued two-dimensional tetrahedral sheet and broken octahedral sheets. Their form can be defined as ribbons of 2:1 phyllsilicate structure. Each ribbon is bonded to the next by inversion of SiO_4 tetrahedra along a set of Si-O-Si bonds. Channels between the ribbons are generated due to these structural properties, which are larger in Sep (4 Å x 9.5 Å) in comparison with Pal (4 Å x 6 Å). Pal is dioctahedral and Sep is trioctahedral. Both minerals have elongated habits often forming bundles of lath-like or fibrous crystals. They contain two types of water, structural water coordinated to the octahedral cations and zeolitic water, which is loosely bound in the channels. Due to the existence of channels, Pal and Sep have great microporous volumes. They are also of a big interest and importance due to their large specific surface, exceptional swelling, and good absorbability [25, 38].

2.3 Clay minerals modification

Modification of clay minerals' native physicochemical and structural characteristics (porosity, CEC, acidity, surface area, etc.) is usually required to adapt and extend their use in different fields of application [20].

Several physical and chemical methods have been investigated to modify clay minerals features, such as acid activation, intercalation or adsorption of organic components, clay pillarization, synthesis of porous clays heterostructures, ultrasonication, and thermal treatment [39–42].

3. Essential oil: chemical composition and properties

Essential oils (EOs) are volatile liquids extracted from numerous plants. They are hydrophobic and contain volatile aromatic compounds or vegetable essences. These aromatic compounds are volatile organic compounds (VOCs), which result from the plant secondary metabolism [43]. According to the European Pharmacopoeia, an essential oil is defined as an "odorous product, commonly with a complex composition, obtained from a plant raw material by steam distillation, dry distillation, Hydrodistillation or other suitable methods without heating" [44].

Essential oils could be obtained from flowers (Pelargonium Rosat, Lavanduladentata), leaves (Eucalyptus globulus, Thymus vulgaris, Origanum vulgare, melaleuca alternifolia), seeds (Coriandrumsativum, Carumcarvi, Foeniculum vulgare, Pimpinellaanisum), peel (Citrus sinensis, CitrusLimon), rhizomes (Zingiberofficinale), and woods (Cinnamomum Cassia, Santalum album) [45].

Generally, EOs present less than 5% of the vegetal dry matter. Their composition may vary depending on the plant organ utilized, the soil and climatic conditions, and the harvesting season. Essential oils are insoluble in inorganic solvents, while soluble in organic ones (alcohol, ether, and oils). They are colorless, liquid, and volatile at room temperature, having a characteristic odor with a density less than water density [8].

EOs are very interesting natural products with several biological properties, such as antioxidative, anti-inflammatory, cytoprotective, antitumor, antimicrobial, antihypertensive, analgesic, larvicidal, insecticidal, antiparasitic, and other biological activities [5]. Thus, they have been largely used in agriculture, pharmaceutical field, medicine, and cosmetics. Indeed, numerous studies have shown insect-repellant and biocide activities, which are usable in agronomy and food industry [46, 47]. Due to their incredible properties, interest in EOs has massively increased in recent years. They are subsequently useful as complementary medicinal treatments, due to their availability and to synergistic therapeutic effects with conventional medicines, such as antibiotics [48]. Aspects related to the significant antimicrobial effect against multiresistant pathogens of the EOs, and their synergetic effect when one or more oils are mixed used in combination with known drugs [49, 50], are also presented in a recent review [51].

3.1 Chemical composition of essential oils

The number of known compounds present in essential oils has recently increased with the improvement in instrumental analytical chemistry. Nowadays, there are more than 300 components present in pure EOs [52, 53] that can be classified into two categories:

- Volatile compounds: Volatile fraction presents about 90% of total oil weight. It comprises terpenes and their oxygenated derivatives named terpenoids. Aldehydes, esters, and aliphatic alcohols may also be present in volatile fraction.

- Nonvolatile compounds: Nonvolatile residue, which presents less than 10% of total EO in weight, contains hydrocarbons, sterols, waxes, carotenoids, flavonoids, and fatty acids [54].

As mentioned before, the main component of the chemical composition of EOs is a complex mixture of hydrocarbon terpenes and terpenoids. Obviously, chemical compounds contain carbon and hydrogen as their building blocks that are isoprene (C5H8), the basic hydrocarbon unit found in EOs [46]. Moreover, terpenes are a class of natural products represented by general structural formula (C5H8)n. Containing more than 30.000 compounds, these unsaturated hydrocarbons are developed by plants, particularly conifers. The classification of terpenes is based on the number of isoprene units: Two, three, or four isoprene units are joined head to tail form monoterpenes, sesquiterpene, and diterpenes, respectively. Monoterpenes are present with more than 80% of EOs composition with 10 carbons, while sesquiterpenes and diterpenes are composed of 15 and 20 carbons, respectively. However, hemiterpenes (C5), diterpenes (C20), triterpenes (C30), and tetraterpenes (C40) are also observed in nature [55–57].

The second class of terpenes containing oxygen is called terpenoids. Those oxygenated derivatives of hydrocarbon terpenes can be aldehydes, alcohols, esters, ketones, acids, and phenols. Terpenoids also present many biological properties such as anti-inflammatory, antifungal, antiseptic, bactericidal, antiviral, sedative, and antitumor [58–60].

Furthermore, non-terpene components, defined as phenylpropanoids, are also found in EOs providing odor and a particular flavor (Eugenol and Cinnamaldehyde) [55].

3.2 biological properties

EOs have, for a long time, been recognized to possess several biological actions on humans, animals, and plants, namely [61–65]:

- Antioxidant activity
- Anticancer activity
- Antimicrobial activity
- Antiviral activity
- Antiparasitic activity
- Anti-inflammatory activity
- Insecticidal activity

4. Application of clay-essential oil nanohybrids

Nanostructured materials hailed from clays and EOs are increasingly attractive for diverse applications, due to their eco-friendly nature and hybrid character that affords original features to the implied organic and inorganic components.

4.1 Active packaging and coatings

Active packaging is defined as the packaging technology in which active agents are incorporated to the original packaging system in order to inhibit food contamination and oxidation, maintain the quality and the safety of food longer, and prolong shelf life by preserving food from internal and external environmental factors [66, 67].

Synthetic additives with bioactive properties have been amply used in food industry. However, growing health and ecological concerns due to the use of chemical food ingredients have amplified the consumer interest toward natural bioactive substances such as essential oils generally recognized as safe (GRAS) by the Food and Drug Administration of the United States and thus are promising alternatives to synthetic additives [68].

However, their high volatility, low solubility in water, sensitivity for oxidation, low photothermal resistance, and undesirable flavor restrain their use [69, 70].

In order to overcome these barriers, EOs' encapsulation and immobilization may be considered as an interesting solution to decrease these drawbacks and to develop new practical antimicrobial packaging materials [71, 72]. Based on their abundance, bio-inert nature, unique layered structure, intercalation, and swelling properties and high retention capacities, loading EOs into clays has been proved to be appropriate for protecting and preserving the efficiency of EOs in storage, providing a controlled release of EOs in polymer matrix. Studies about clay/EOs nanohybrid application in active packaging sector are still emerging.

The encapsulation of thyme essential oil (TO) into halloysite nanotubes, *via* a vacuum mode, for application in the food packaging, was studied by Lee and Park [73]. TO-loaded HNTs capsules coated with Eudragit polymer (EPO) were also prepared in the purpose to prevent burst release and to extend the release time. The amount of TO released from the HNT/TO and EPO/HNT/TO systems was 61.76% and 45.27% at 24 h, respectively. Also, it was observed that the TO liberation from each type of capsule was maintained up 96 h. Additionally, it was noted that TO/HNT hybrids presented an initial burst release of 47.96% within 12 h, while EPO/HNT/TO hybrids exhibited important retarded TO release. Finally, it was concluded that HNT capsules containing TO exposed good antioxidant activity, compared with Pristine HNTs that showed no antioxidant activity.

In order to increase the loading capacity of HNTs, another research work was investigated by Lee et al. [74] consisting of developing an antimicrobial nanosystem based on TO and modified HNTs (alkali-HNT) by treating pristine HNTs with sodium hydroxide (NaOH, 5 mol/L). It was demonstrated that the encapsulation efficiency passed from 14.5% to 20.5% for raw HNT clays and treated HNT clays, respectively. Moreover, it was found that the release rate of TO in liquid form was very fast than that of TO in hybrid form, which suggests that TO was strongly retained by HNTs in the original and treated state.

The inactivation of *E. coli*, total mesophilic zerobic bacteria (MAB), and molds and yeasts (MY) on the surface of the cherry tomatoes exposed to alkali-HNT/TO hybrids was assessed. It was deduced that the populations of *E. coli*, MAB, and MY could be reduced over the storage time of fresh cherry tomatoes after the addition of alkali-HNT as a nanodelivery system for the controlled aerial release of TO.

A third study dealing with the encapsulation of TO in HNTs was conducted by Jang et al. [75]. The TO was loaded in HNTs *via* vacuum pulling methods, followed by end-capping or a layer-by-layer surface coating process for complete loading. After, the nanohybrids were combined with flexographic ink and covered on a food packaging paper.

Encouraging results were emanated from the study of the EO release and the packaging paper antibacterial activity and emphasized the interest in using such functional packaging material printed with ink containing TO- and HNTs-based nanocapsules.

HNTs were explored by Biddeci et al. [76] as nanodelivery system for peppermint essential oil (PO). The purpose of the study was to develop an antioxidant/antimicrobial biofilm by filling a pectin matrix with modified HNTs loaded with PO. The modification of HNTs was based on the functionalization of clay surface with cucurbit [6] uril (CB [6]) molecules to increase the affinity of the nanoclay toward PO. Pectin-based biofilms containing HNTs/CB [6]/PO hybrid were prepared by casting method under reduced pressure. A considerable inhibition percentage (41%) for biofilms was evidenced using the antioxidant activity test. While *in vitro* experiments of the antimicrobial properties for *E. coli* and *S. aureus* isolated from beef and cow milk displayed more efficiency at higher temperature.

An investigation dealing with the preparation and characterization of green composites based on pectins and nanohybrids clay/EOs was reported by Gorrasi [77]. HNTs were valued as possible nanocontainers for rosemary essential oil (RO). It was observed that the active agent release from the composite containing nanohybrids was much slower than the release of the same molecule simply added to the matrix. Molds formation was detected in pectin films after 2 weeks of storage at room temperature (25–30°C) and environmental humidity (about 60%). On the contrary, it was not noticed in nanohybrid-based films even after 3 months, suggesting so the promising use of clay/RO nanohybrid in the active packaging field.

A linear low-density polyethylene (LDPE)-based films incorporated with active nanoclay particles were developed adopting a new method for grafting EOs on nanoclay particles, and this by using Tween 80 as surfactant. It was demonstrated that nanoclays permitted a protective effect for the active substances against evaporation during film manufacturing. *In vitro* antibacterial activity of the activated films against pathogenic bacteria (*Salmonella Typhimurium*, *E. coli*, *Listeria monocytogenes*, *S. aureus,* and *Bacillus cereus*) was marked, while their effectiveness against lactic acid bacteria (*Lactobacillus rhamnosus* and Lb. *casei*) was restricted [78].

The formation of new active bilayer films impregnated with attapulgite (ATP) clay nanoparticles loaded with *Allium Sativum* essential oil (AO) and based on LDPE and polypropylene (PP) was investigated using blown film extrusion method. The preservation performances of these potential active films were tested for large yellow croaker at 4°C of storage. It was able to demonstrate that the lipid oxidation of seafood product could be incessantly stopped. This fact was justified by the controlled release of AO from ATP [79].

A recent study was published aiming to explore the efficiency of sodium montmorillonite (Na-MMT) and organically modified Montmorillonite (Org-MMT) to be nanodelivery systems for the controlled release of EOs in LDPE active films. Thyme, oregano, and basil EOs were chosen for their antioxidant property. It was found that the antioxidant activity of films varied depending on the EO type and content. It was noted also that the use of Na-MMT and Org-MMT as nanocontainers for EOs allowed managing the antioxidant activity of the elaborated films [80].

The formulation of a sustained liberation antibacterial chitosan (CS)-based packaging film, by casting solution method, through filling CS matrix by biological active nanoparticles was presented by Cui et al. [81]. The active nanofillers resulted from the loading of cinnamaldehyde (CIN), the major constituent of cinnamon essential oil, into acid-treated montmorillonite (acid-MMT). The search in the CIN release rate from CS films showed stable rates for CS/CIN and CS/acid-MMT/CIN films equal to 65.01% and 73.20% at 96 h and 168 h, respectively. These outcomes emphasized the promising use of acid-MMT nanoclay as nanoencapsulation materials for slow release of EOs components. *In vitro* antibacterial activity of acid-MMT/CIN-based biofilm was tested for the growth of *S. aureus* and *E. coli*. Preliminary results displayed a noticeable inhibitory effect.

A new second study was also published by Cui et al. [82] dealing with the development of sodium alginate (SA)-based active package with controlled release of CIN loaded into HNTs. To enhance the uptake capacity of HNTs, the clay nanoparticles were treated with sulfuric acid (acid-HNT). Analog results were found for SA/acid-HNT/CIN film for the slow release behavior of the CIN and *in vitro* antibacterial activity compared with CS/acid-MMT/CIN film.

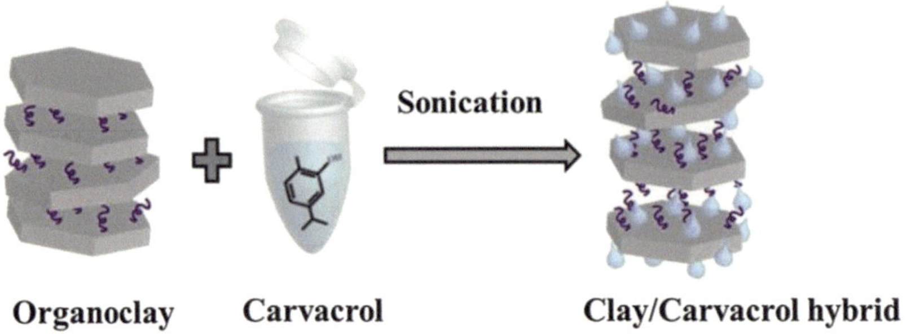

Figure 2.
A schematic illustration of organoclay galleries modified with carvacrol molecules as achieved by a pre-compounding step in which clay/carvacrol hybrids are produced.

Recently, lemon waste natural dye (LD) and EO (LO) were valued in the hybrid form with MMT laminar nanoclays, for a potential application as nanofiller for polyester based matrix. The co-adsorption of LD and LO on MMT nanoparticles was optimized by using statically designed experiments. The polyester-based bionanocomposites were prepared for different nanohybrid loading rates (3, 5, and 7 wt %). The experiments carried out in this study were devoted to assess the effect of nanohybrid (MMT/LD/LO) incorporation on thermomechanical and color properties of the polymer matrix [83].

LDPE/clay nanocomposites films, comprising carvacrol (CRV) with controlled and tunable antimicrobial activity, were conceived. Org-MMT/CRV hybrids were prepared by shear mixing CRV with org-MMT at a weight ratio 2:1 (respectively) followed by ultrasonication at room temperature for 20 min at constant amplitude of 40% (**Figure 2**) [84].

After melt compounding and compression molding, LDPE/org-MMT/CRV films presented higher CRV content (5–6 wt %) in comparison with control LDPE/CRV films, containing only ~ 3 wt%. As follows, it was concluded that clay platelets acted as nanocontainers for volatile compounds of CRV while improving their thermal stability during high-temperature process.

The antimicrobial activity of the films against *E. coli* bacteria was assessed. It was found that freshly prepared films reduced *E. coli* cells to an undetectable rate, affirming the effective bactericidal activity of CRV within the melt-compounded films. But, 1-month-old LDPE/org-MMT/CRV films, stored at room temperature, showed lower antimicrobial potency. However, the LDPE/CRV films totally loosed their biological potency.

An analog comportment of the tested films was noticed against Gram-positive (*Listeria innocua*) bacteria in a second study conducted by Shemesh et al. [85] where it was also described and discussed the excellent antifungal activity of LDPE/org-MMT/CRV films against the phytopathogenic and clinical fungus *A. alternata*.

Another study was exposed by Shemesh et al. [86], whose objective was to investigate the use of HNTs as nanocarriers for CRV for its later melt compounding with LDPE. Like it was demonstrated for LDPE/org-MMT/CRV nanocomposites films, HNTs showed their vital role in improving CRV thermal properties during LDPE/HNTs/CRV films manufacturing, as well as controlling and delaying the CRV release.

The co-encapsulation of CRV and thymol (TYM) into HNTs for developing active food packaging film with synergistic antimicrobial activity was examined by Krepker et al. [87]. Satisfactory results were achieved and the resulting films revealed superior antimicrobial activity against *E. coli* when compared with LDPE based films including the individual EOs. This was attributed to the synergistic interactions between CRV and TYM.

Rosewood, manuka, oregano, and lavender EOs were valued, by Kinninmonth et al. [88], by their adsorption onto natural and acid-treated bentonites for their controlled release and protection against polymer processing conditions. Promising results were reported.

A smart waterborne paint was designed by incorporating bioactive hybrids in the formulation. The active nanosystems were conceived based on org-MMT, Na-MMT, and citronellol (CIT).

The antimicrobial activity of the synthesized hybrids was assessed against *Chaetomium globosum* and *Alternaria alternate*. The outcomes were promising and revealed that org-MMT-based hybrid was more active than Na-MMT-based hybrid.

The bio-resistance tests were conducted, on the acrylic paint indoor formulation containing the org-MMT/CIT nanohybrid and that containing org-MMT and CIT added independently, by exposing the films during 4 weeks to fungal growth in plates. It was observed that the paint including org-MMT/CIT hybrid exposed no growth. By contrast, the paint containing free CIT showed abundant fungal growth. This could be explained by the volatility of CIT when it is added in free form in the paint formulation, and therefore, it would not be available to afford its antifungal activity. Additionally, the integration of CIT in a free form could conduct to reactions with some added ingredients in the paint formulation, which would also affect its bioactivity. On another side, the excellent achievement obtained with coatings comprising the nanohybrid material could be correlated with the sustained and controlled release of CIT from org-MMT [89].

4.2 Bio-derived agrochemicals

Agrochemicals are crucial ingredients needed to reach general food security. About 2.5 million tons of synthetic agrochemicals, comprising fertilizers, pesticides, herbicides, fungicides, insecticides, and others, are used each year [90]. The overuse of these kinds of compounds leads to a major pollution for both soil and water, with a high toxicity toward humans and animals. EOs are an excellent substitute to synthetic agrochemicals as a way to lower negative impacts to human health and the environment. Nevertheless, some of their properties, *for example,* water insolubility, chemical instability, degradation by temperature and light, may be hindrances to their use as biocontrol agents [91, 92]. Clay nanoparticles have been widely studied as nanocarriers for synthetic agrochemicals over the last decade [93–96]. Moreover, studies dealing with nanoclay/bio-derived agrochemicals-based formulations are emerging.

An insecticidal powder formulation based on *Ocimum gratissimum* EO and org-MMT clay was developed by Nguemtchouin et al. [97]. The insecticidal effect of the nanohybrid org-MMT/O-gratissimum was evaluated against the maize weevil *Sitophilus zeamais*. It was noticed that the mortality of *Sitophilus zeamais* dropped from 100 to 95%, 87%, and 0% after 7 days, respectively, for org-MMT/O-gratissimum, Na-MMT/O-gratissimum, and crude EO. Moreover, organoclay-based formulation was more stable, since it lost about 60% of its full biocidal capacity after 30 days, whereas unmodified clays-based formulation released entirely its insecticidal activity for the same period of time. Experiment conducted for formulation remnant effect determination indicated that the insecticidal effects of O-gratissimum EO remained for about 7, 45, and 80 days, for free EO, EO included in Na-MMT, and EO included in org-MMT, respectively [97].

The efficiency of aromatized powder based on O-gratissimum EO and organically modified clay, as a bioinsecticide for use in pest control of stored maize, was confirmed in a second study reported also by Nguemtchouin et al. [98].

Xylopia aethiopica EO was investigated too by Nguemtchouin et al. [99] for the development of new bioinsecticides based on kaolinite clay. Ingestion-contact insecticide tests were carried out using maize weevil *S. zeamais*. As described in the study, the mortality of *S. zeamais* was proportional to the mass of the powder formulation put together with maize and insects. Insect's mortality varied from 22% with 2.5% (w/w) of bioinsecticide to 100% with 10% (w/w). The remnant effect evaluation showed that the initial mortality rate of insects induced by the clay-based bioinsecticide was the highest. But, with the time it decreased until reaching 0% at the 8th week. This was explained by the volatility of the EO. To overcome this limit, enhancing the clay structural properties by chemical modification was suggested to raise both the adsorption and the retention capacities.

In relation to *X. aethiopica* EO and kaolin-based bioinsecticide formulation, a study was reported on the effect of clay particle size and clay treated with hydrogen peroxide (H_2O_2) on its adsorption capacity. It was observed that the amount of EO adsorbed was inversely proportional to the particle size. The treatment of kaolin with H_2O_2 promoted the adsorption rate of *X. aethiopica* EO components [100].

The adsorption behavior of TYM onto MMT and Kaol clays and their corresponding modified ones (by treatment with aqueous solution of iron polycations) were modelized by Nakhli et al. [101] using a statistical approach to understand this comportment for the future application of these clay samples as adsorbent in biopesticides formulation. Encouraging results were emerged from this study.

In another research paper, published by Nguemtchouin et al. [102], it was exposed the results of the textural and structural modification of bentonite clay using metallic polycations solutions of $Al_x(OH)_y$ and $Fe_x(OH)_y$ to obtain inorganic bentonite and cetyl and phenyl trimethyl ammonium chlorides solutions to obtain organic bentonite. The attempt to adsorb TYM, chosen like an insecticidal terpenic compound, on the different modified bentonites was carried out. Both ways for clay modification were considered as interesting to promote the adsorption properties of clays utilizable as support for bioinsecticides.

Clausena anisata EO was nanoencapsulated in MMT, and the resulting powdery formulation was evaluated by Ndomo et al. [103] for their insecticidal activity and their effects on progeny production of *Acanthoscelides obtectus*. It was pointed out that there was a dose-dependent progress in mortality of *A. obtectus* adults in bean grains treated with clay-EO nanohybrid formulation. It was also highlighted that at the dosage and exposure time, the EO in its nanohybrid form exhibited a higher mortality effects against *A. obtectus* adults than that caused by pure EO applied directly. Finally, it was stated that although EOs applied alone afford an acceptable level of stored grain pests control, they can cause a persistent odor that can be unpleasant when eating the seeds. So, the use of clay as a support material for such compounds could diminish the adverse effects in addition to the increase of their stability.

Essential oils of *Lantana camara L.* (*Verbenaceae*) and *Annona senegalensis Pers.* (*Annonaceae*) were valued by Gueye et al. [104] for their insecticidal effect on adults of *Caryedon serratus*, a pest of groundnut stocks, through using Kaol clay as vehicle for these active agents. Promising outcomes were reported and discussed.

The Kaol clay was also examined by Kéita et al. [105] as a nanocontainer for *Ocimum basilicum L.* and *O. gratissimum L.* and their application as an insecticide in a powder form to control *Callosobruchus maculates*. The effectiveness of the powders formulation was demonstrated and proved by varied bioassays.

A new approach for clay and EO-based bioinsecticide formulation was exposed by Noudem et al. [106]. It considered the modification of MMT by saponins (Sap-MMT) to enhance its adsorption capacity toward *O. gratissimum* EO compounds. The outcomes of the formulations remnant effect tested against *Callosobruchus subinnotatus* evidenced the decrease of the insects' mortality from 96 to 70% and from 96% to 13.12% for Sap-MMT/*O. gratissimum* nanosystem and Na-MMT/*O. gratissimum* nanosystem after 42 days of exposure, respectively. This finding was explained by the interaction types existing between the terpenic compounds of the EO and Na-MMT or Sap-MMT. Additionally, the efficiency of Sap-MMT/*O. gratissimum* nanosystem could be also owing to the high quantity of terpenic compounds adsorbed by Sap-MMT compared with Na-MMT.

Adsorption of EO components of *Lavandula angustifolia* and TYM on Moroccan-modified bentonite, for a potential use, respectively, as an insecticide and an acaricide, was explored by El Miz et al. [107–109].

Organically modified palygorskite and beidellite clays were explored by Ghrab et al. [110, 111] as support materials for *Eucalyptus globulus* EO-active terpenic compounds adsorption for a potential insecticidal application.

The adsorption of *Lippia Multiflora* EO on two organically modified clays, Kaolinite-rich clay and smectite-rich clay, was explored *via* a qualitative approach using X-ray diffraction and infrared analyses. It was deduced from this investigation that smectite-rich clay can be considered as a suitable material for producing biopesticides compared with kaolinite-rich clay [112].

TYM was immobilized by Ziyat et al. [113] onto purified Moroccan clay "Rhassoul" and the organically modified one. It was estimated that the adsorption capacities were of 6 mg/g and 16 mg/g for purified Rhassoul and modified one, respectively. Hence, organo-modified Rhassoul was regarded as an attractive material for biofungicides formulation based on EO.

Recently, a study was focused on the evaluation of the antifungal activity of thyme and oregano EOs combined with purified Rhassoul and sulfuric-acid-activated Rhassoul. *Penicillium sp*. was chosen as a pathogen agent for biological activity. *In vitro* tests demonstrated that activated-clay-based formulations exhibiter higher inhibition power than purified clay-based formulations. This was attributed, in part, to the adsorbed quantity by activated clay, which was higher than that retained by purified one. Nevertheless, it was considered that EOs in a nanohybrid form with Rhassoul could be applied as an alternative to synthetic fungicides to prevent fungal growth during grain storage [114].

CRV/ATP nanohybrid antibacterial materials were designed by Zhong et al. [115] using a grinding process in the intention to replace the synthetic antibiotic used in animal farming. The antibacterial activity of the hybrid was evaluated against *E. coli* and *S. aureus*, and the minimum inhibitory concentration was equal to 2.0 mg/mL for both bacteria model.

A recent study was also published by Zhong et al. [116] reporting the preparation of a series of antibacterial hybrid materials based on Pal and different EOs.

CRV was also valued by Berraaouan et al. [117] by hybridization with purified bentonite for application in pesticide industry and in other fields. The adsorption studies revealed attractive results.

A series of nanohybrid materials were prepared by adsorbing natural active components, such as eugenol, CIN, TYM, allyl isothiocyanate, and diallyl disulfide onto MMT. The biological activity of the nanomaterials was assessed against *S. aureus* and *Aspergillus niger* (*A. niger*). It was observed that the nanoencapsulation enhanced both the antifungal and antibacterial activities of the EOs components. Thereby, the developed nanocapsules may find uses in integrated pest management systems in organic agriculture. The bioactive substances loading into nanoclay may ensure their controlled release in accordance with the needs of vegetation while simultaneously helping to reduce environmental pollution [118].

Newly, a study was conducted by Saucedo-Zuñiga et al. [119] dealing with the preparation and the characterization of a multilayer film reservoir including EOs/clay nanohybrids for a potential application as pesticide or attractant for pest control as well as coatings for antimicrobial or fungicidal control. Two kinds of nanoclays were investigated, HNTs and org-MMT, as well as two kinds of EOs, thyme and orange EOs. Promising results were reported and discussed for the convincing use of this multilayer film encapsulated EO/clay composites as aroma-controlled release systems for pesticides and for also active food packaging applications.

More recently, a new green evaporation/adsorption method was exposed by Essifi et al. [120] for the adsorption of EOs on Na-MMT. The aim of the preparation process was to elaborate powdered EO/Na-MMT hybrids for acaricidal, fungicidal, larvicidal, and insecticidal applications. Promising results were reported for the potential use of this kind of nanomaterials.

4.3 Pharmaceutical and cosmetic applications

Hybrid structures based on clay and synthetic drug for biomedical and pharmaceutical applications have been amply investigated over the past decade. Nowadays, reports treating clay/natural bioactive substances nanohybrids are arising [20].

A bio-based antimicrobial mosquito repellent was developed by immobilizing a mixture of Curmuma aromatic and Zanthoxylum limonella EOs onto MMT clays dispersed in methyl ester of the castor oil. The estimation of the mosquitocidal activity and the antibacterial activity of the resulting nanosystems revealed that the formulation efficiency is EOs ratio-dependent [121].

The attempt to develop a gelling viscoelastic film loaded with CRV/clay nanohybrid for infected skin ulcer treatment was exposed by Tenci et al. [122]. MMT, HNTs, and Pal clays were examined for the nanosystems preparation. It was reported that CRV loading capacity was clay structure dependent. The highest amount was obtained for, the fibrous clay, Pal. *In vitro* assessment of cytocompatibility, and antioxidant and antimicrobial properties was conducted for pure CRV and its corresponding optimal nanohybrid (CRV/ Pal). It was marked that CRV displayed a cytotoxic effect for concentrations >50 μg/mL. However, the corresponding nanohybrid did not present any cytotoxic effect for all the tested concentrations. According to the authors, the incorporation of CRV into Pal permitted to keep safe human fibroblasts against CRV cytotoxicity owing to a controlled release of CRV to cell membrane. The efficiency of CRV in protecting human fibroblasts against oxidative stress was confirmed by the antioxidant properties determination. Even its hybridization with Pal did not affect such activity. The antimicrobial activity experiments tested against *S. aureus* and *E. coli* proved a decrease of the minimum inhibitory concentration and the minimum bactericidal concentration for CRV loaded into Pal compared with the pure one, meaning that CRV in the hybrid form manifested a higher antimicrobial activity.

This observation was interpreted by the low evaporation of CRV after loading into clay and its slow release from the nanostructure.

Nanohybrid materials based on limonene and lecithins-modified MMT were invented by Nagy et al. [123] to be applied as flavor and fragrance nanodelivery system.

A new academic work was reported looking at the formulation of novel nanodelivery systems with CRV prodrugs and fibrous clays, namely Pal and Sep in the purpose to restrict the chemical conversion of the loaded CRV prodrugs to CRV in the distal small intestine where EOs can execute its maximum antimicrobial activity. Pal was found to have a better affinity to CRV prodrugs than Sep. *In vitro* release profiles were investigated under conditions simulating the gastric and intestinal transit. It was concluded that tested clays were able to ensure a continuous release of CRV prodrugs that are subjected to delayed conversion into antimicrobial active CRV as demanded to consolidate its pharmacological activity [124].

A nanohybrid based on PAL and ginger EO (GO) was developed by Lei et al. [125]. The EO is known for its pharmacological properties such as antibacterial, antioxidant, and antinociceptive. The antibacterial activity of PAL/GO tested against *E. coli* and *S. aureus* revealed minimum inhibitory concentration (MIC) values for GO in PAL/GO hybrid equal to 1.16625 mg/mL and 4.665 mg/mL against *S. aureus* and *E. coli*, respectively, while free GO presented MIC values equivalent to 3.5625 mg/mL against *S. aureus* and 7.125 mg/mL against *E. coli*. These findings confirmed the high antibacterial activity of GO in hybrid form with PAL compared with its free form. The thermo-stability and acidity and alkalinity-resistance tests were exploited to demonstrate that PAL/GO nanosystems conserve their antibacterial activity in extreme conditions.

Natural MMT and Tween 20-modified MMT were employed as nanocarriers for cinnamic acid delivery systems for oral administration. It was underlined the interest in surfactant using for clay-based nanocarrier preparation, as it ensured the complete release of the cinnamic acid after oral drug administration, in addition to the improvement of the drug loading [126].

5. Conclusion

Overall, the hybrid system obtained from nanoclays with EOs is an efficient approach to protect EOs from light, air, and humidity, which lead to oxidation or volatilization of EOs and a reduction of their biological properties. Moreover, the hybrid system increases the solubility and physicochemical stability of EOs and offers a controlled and sustained release rate of EOs and makes them more available. The hybrid nanoclay/EOs open new perspectives in cosmetic, food, and pharmaceutical industries and could be an economic benefit as it is inexpensive and also fulfills the consumer concern regarding safety as it is environmentally friendly and nontoxic agent.

Author details

Houda Saad[1*], Ameni Ayed[2], Mondher Srasra[1], Sameh Attia[1], Ezzeddine Srasra[1], Fatima Charrier-El Bouhtoury[3] and Olfa Tabbene[2]

1 Laboratory of Composite Materials and Clay Minerals (LMCMA), National Center of Researches in Material Sciences (CNRSM), Soliman, Tunisia

2 Laboratory of Bioactive Substances, Center of Biotechnology of Borj-Cedria (CBBC), Hammam-lif, Tunisia

3 IPREM – UMR CNRS 5254, IUT des Pays de l'Adour, University of Pau and Pays of l'Adour (UPPA), Mont de Marsan, France

*Address all correspondence to: saad_houda@yahoo.com

References

[1] Tiwari S, Singh BK, Dubey NK. Encapsulation of essential oils - A booster to enhance their bio-efficacy as botanical preservatives. Journal of Scientific Research. 2020;**64**:175-178. DOI: 10.37398/JSR.2020.640125

[2] Majeed H, Bian YY, Ali B, Jamil A, Majeed U, Khan QF, et al. Essential oil encapsulations: Uses, procedures, and trends. RSC Advances. 2015;**5**:58449-58463. DOI: 10.1039/C5RA06556A

[3] Radünz M, Helbig E, Borges CD, Gandra TKV, Gandra EA. A mini-review on encapsulation of essential oils. Journal of Analytical & Pharmaceutical Research. 2018;**7**:00205. DOI: 10.15406/japlr.2018.07.00205

[4] DajicStevanovic Z, Sieniawska E, Glowniak K, Obradovic N, Pajic-Lijakovic I. Natural macromolecules as carriers for essential oils: From extraction to biomedical application. Frontiers in Bioengineering and Biotechnology. 2020;**8**:563. DOI: 10.3389/fbioe.2020.00563

[5] Chiriac AP, Rusu AG, Nita LE, Chiriac VM, Neamtu I, Sandu A. Polymeric carriers designed for encapsulation of essential oils with biological activity. Pharmaceutics. 2021;**13**:631. DOI: 10.3390/pharmaceutics13050631

[6] Bakry AM, Abbas S, Ali B, Majeed H, Abouelwafa MY, Mousa A, et al. Microencapsulation of oils: A comprehensive review of benefits, techniques, and applications. Comprehensive Reviews in Food Science and Food Safety. 2016;**15**:143-182. DOI: 10.1111/1541-4337.12179

[7] Maes C, Bouquillon S, Fauconnier ML. Encapsulation of essential oils for the development of biosourced pesticides with controlled release: A review. Molecules. 2019;**24**:2539. DOI: 10.3390/molecules24142539

[8] Lammari N, Louaer O, Meniai AH, Elaissari A. Encapsulation of essential oils via nanoprecipitation process: Overview, progress, challenges and prospects. Pharmaceutics. 2020;**12**:431. DOI: 10.3390/pharmaceutics12050431

[9] Donsì F, Annunziata M, Sessa M, Ferrari G. Nanoencapsulation of essential oils to enhance their antimicrobial activity in foods. LWT - Food Science and Technology. 2011;**44**:19081914. DOI: 10.1016/j.lwt.2011.03.003

[10] Nanoclays YR. A new avenue for drug delivery. EC Pharmacology and Toxicology. 2019;**ECO.02**:20-22

[11] Floody MC, Theng BKG, Reyes P, Mora ML. Natural nanoclays: Applications and future trends _ a Chilean perspective. Clay Minerals. 2009;**44**:161-176. DOI: 10.1180/claymin.2009.044.2.161

[12] Fossum JO. Clay nanolayer encapsulation, evolving from origins of life to future technologies. The European Physical Journal. 2020;**229**:2863-2873. DOI: 10.1140/epjst/e2020-000131-1

[13] Murray HH. Overview - Clay mineral applications. Applied Clay Science. 199;**15**:379-395. DOI: 10.1016/0169-1317(91)90014-Z

[14] Fabrícia de Castro S, Luciano ClécioBrandão L, Luzia Maria Castro H, Pollyana T, JosyAnteveli O, Anderson Oliveira L, et al. Clays as biomaterials in controlled drug release: A scientific and technological short review. Biomedical Journal of Scientific and

Technical Research. 2019;**15**:11237-11242. DOI: 10.26717/BJSTR.2019.15.002677

[15] Jlassi K, Krupa I, Chehimi MM. Overview: Clay preparation, properties, modification. In: Deans M, editor. Clay-Polymer Nanocomposites. Cambridge, United States: Elsevier; 2017. pp. 1-28. DOI: 10.1016/B978-0-323-46153-5.00001-X

[16] Gürses A. Introduction to Polymer-Clay Nanocomposites. 1st ed. New York: Pan Stanford Publishing Pte. Ltd; 2015. p. 88

[17] Bergaya F, Lagaly G. General introduction: Clays, clay minerals, and clay science. In: Bergaya F, Theng BKG, Lagaly G, editors. Handbook of Clay Science. Amsterdam, Netherlands: Elsevier; 2006. pp. 1-18. DOI: 10.1016/S1572-4352(05)01001-9

[18] Utraki AL. Clay-containing Polymeric Nanocomposites. Vol. 1. Shawbury, Shrewsbury, Shropshire, UK: iSmithersRapra Publishing; 2004. p. 785

[19] White RE. Principles and Practice of Soil Science, the Soil as a Natural Resource. 4th ed. Vol. 384. Singapore: Wiley-Blackwell; 2005

[20] Saad H, de Hoyos-Martinez PL, Srasra E, Charrier-El BF. Advances in bio-nanohybrid materials. In: Ahmed S, Haussain CM, editors. Green and Sustainable Advanced Materials. NJ, USA: Scrivener Publishing LLC, Wiley; 2018. p. 378

[21] Barakan S, Aghazadeh V. A review on the advantages of clay mineral modification methods for enhancing adsorption efficiency in wastewater treatment. Environmental Science and Pollution Research. 2020;**28**:2572-2599. DOI: 10.1007/s11356-020-10985-9

[22] Obaje SO, Omada JI, Dambatta UA. A review on clays and their industrial applications: Synoptic review. International Journal of Science and Technology. 2013;**3**:264. DOI: 10.30564/re.v3i2.3057

[23] Kumari N, Mohan C. Basics of Clay Minerals and Their Characteristic Properties. London: IntechOpen; 2021. Available from: https://www.intechopen.com/online-first/76780 [Accessed: August 9, 2021]

[24] Nanzyo M, Kanno H. Secondary minerals. In: Inorganic Constituents in Soil. 1st ed. Singapore: Springer; 2018. p. 181. DOI: 10.1007/978-981-13-1214-4

[25] Christidis GE. Industrial clays. In: Christidis GE, editor. Advances in the Characterization of Industrial Minerals. Mineralogical Society of Great Britain and Ireland; 2011. pp. 341-414. DOI: 10.1180/EMU-notes.9

[26] Li S, Mu B, Wang X, Kang Y, Wang A. A comparative study on color stability of anthocyanin hybrid pigments derived from 1D and 2D clay minerals. Materials (Basel). 2019;**12**:3287. DOI: 10.3390/ma12203287

[27] Yuan P, Tan D, Annabi-Bergay F. Properties and applications of halloysite nanotubes: Recent research advances and future prospects. Applied Clay Science. 2015;**112**:75-93. DOI: 10.1016/j.clay.2015.05.001

[28] Satish S, Tharmavaram M, Rawtani D. Halloysite nanotubes as a nature's boon for biomedical applications. Nanobiomedicine. 2018;**6**:1-16. DOI: 10.1177/1849543519863625

[29] Brigatti M, Galán E, Theng B. Structures and mineralogy of clay minerals. In: Bergaya F, Theng BKG, Lagaly G, editors. Handbook of Clay

Science. Amsterdam: Elsevier; 2006. pp. 19-86

[30] Du M, Guo B, Jia D. A review on newly emerging applications of halloysite nanotubes. Polymer International. 2010;**59**:574. DOI: 10.1002/pi.2754

[31] Lvov Y, Aerov A, Fakhrullin R. Clay nanotube encapsulation for functional biocomposites. Advances in Colloid and Interface Science. 2014;**207**:189-198. DOI: 10.1016/j.cis.2013.10.006

[32] Ray SS, Okamoto M. Polymer/layered silicate nanocomposites: A review from preparation to processing. Progress in Polymer Science. 2003;**28**:1539-1641. DOI: 10.1016/j.progpolymsci.2003.08.002

[33] García-Romero E, Suárez M. Structure-based argument for non-classical crystal growth in natural clay minerals. Mineralogical Magazine. 2018;**82**:171-180. DOI: 10.1180/minmag.2017.081.031

[34] Awad ME, López-Galindo A, Setti M, El-Rahmany MM, Iborra CV. Kaolinite in pharmaceutics and biomedicine. International Journal of Pharmaceutics. 2017;**533**:34-48. DOI: 10.1016/j.ijpharm.2017.09.056

[35] Schulze DG. Clay minerals. In: Hillel D, Hatfield JL, editors. Encyclopedia of Soils in the Environment. Vol. 2. Amsterdam, Netherlands: Elsevier; 2011. pp. 246-254

[36] Awad ME, López-Galindo A, Setti M, El-Rahmany MM, Iborra CV. Kaolinite in pharmaceutics and biomedicine. International Journal of Pharmaceutics. 2017;**533**:34-48. DOI: 10.1016/j.ijpharm.2017.09.056

[37] Park JH, Shin HJ, Kim MH, Kim JS, Kang N, Lee JY, et al. Application of montmorillonite in bentonite as a pharmaceutical excipient in drug delivery systems. Journal of Pharmaceutical Investigation. 2016;**46**:363-375. DOI: 10.1007/s40005-016-0258-8

[38] Galan E. Properties and applications of palygorskite-sepiolite clays. Clay Minerals. 1996;**31**:443

[39] Bergaya F, Lagaly G. Surface modification of clay minerals. Applied Clay Science. 2001;**19**:1-3. DOI: 10.1016/S0169-1317(01)00063-1

[40] Cecilia J, García-Sancho C, Vilarrasa-García E, Jiménez-Jiménez J, Rodriguez CE. Synthesis, characterization, uses and applications of porous clays heterostructures: A review. The Chemical Record. 2018;**18**:1085-1104. DOI: 10.1002/tcr.201700107

[41] Lagaly G, Ogawa M, Dékány I. Clay mineral organic interactions. In: Bergaya F, Lagaly G, editors. Handbook of Clay Science. Amsterdam: Elsevier; 2013. pp. 309-377

[42] Bhattacharyya KG, Gupta SS. Adsorption of a few heavy metals on natural and modified kaolinite and montmorillonite: A review. Advances in Colloid and Interface Science. 2008;**140**:114-131. DOI: 10.1016/j.cis.2007.12.008

[43] Baser KHC, Buchbauer G. In: Husnu Can Baser K, Buchbauer G, editors. Handbook of Essential Oils: Science, Technology, and Applications. 2nd ed. Boca Raton, FL, USA: CRC Press; 2015

[44] European Pharmacopeia, editor. European Directorate for the Quality of Medicines and Health Care. 9th ed. European Pharmacopeia: Strasbourg, France; 2018

[45] Dhifi W, Bellili S, Jazi S, Bahloul N, Mnif W. Essential oils' chemical characterization and investigation of some biological activities: A critical review. Medicines (Basel). 2016;**3**:25. DOI: 10.3390/medicines3040025

[46] El Asbahani A, Miladi K, Badri W, Sala M, Addi EHA, Casabianca H, et al. Essential oils: From extraction to encapsulation. International Journal of Pharmaceutics. 2015;**483**:220-243. DOI: 10.1016/j.ijpharm.2014.12.069

[47] Obolskiy D, Pischel I, Feistel B, Glotov N, Heinrich M. Artemisia dracunculus L. (tarragon): A critical review of its traditional use, chemical composition, pharmacology, and safety. Journal of Agricultural and Food Chemistry. 2011;**59**:11367-11384. DOI: 10.1021/jf202277w

[48] Stea S, Beraudi A, De Pasquale D. Essential oils for complementary treatment of surgical patients: State of the art. Evidence-Based Complementary and Alternative Medicine. 2014;**2014**:726341. DOI: 10.1155/2014/726341

[49] Padalia H, Moteriya P, Baravalia Y, Chanda S. Antimicrobial and synergistic effects of some essential oils to fight against microbial pathogens: A review. In: Méndez-Vilas A, editor. The Battle Against Microbial Pathogens: Basic Science, Technological Advances and Educational Programs. 1st ed. Spain: Formatex Research Center S.L; 2015. pp. 34-45

[50] Hemaiswarya S, Kruthiventi AK, Doble M. Synergism between natural products and antibiotics against infectious diseases. Phytomedicine. 2008;**15**:639-652. DOI: 10.1016/j.phymed.2008.06.008

[51] Rai M, Paralikar P, Jogee P, Agarkar G, Ingle AP, Derita M, et al. Synergistic antimicrobial potential of essential oils in combination with nanoparticles: Emerging trends and future perspectives. International Journal of Pharmaceutics. 2017;**519**:67-78. DOI: 10.1016/j.ijpharm.2017.01.013

[52] Hanif MA, Nisar S, Khan GS, Mushtaq Z, Zubair M. Chapter 1 Essential Oils. In: Malik S, editor. Essential Oil Research. Switzerland AG: Springer Nature; 2019. pp. 4-10

[53] Zellner BDA, Dugo P, Dugo G, Mondello L. Analysis of essential oils. In: Handbook of Essential Oils. London: CRC Press. Taylor and Francis Group; 2010. pp. 151-184

[54] Rao VPS, Pandey D. A project report on Extraction of essential oil and its applications for Bachelor of Technology (Chemical Engineering) at Department of Chemical Engineering National Institute of Technology Rourkela-769008 Orissa, India, 2007

[55] Moghaddam M, Mehdizadeh L. Chemistry of essential oils and factors influencing their constituents. In: Grumezescu AM, Holban AM, editors. Soft Chemistry and Food Fermentation. Academic Press: London, UK; 2017. pp. 379-419

[56] Paduch R, Kandefer-Szerszeń M, Trytek M, et al. Terpenes: Substances useful in human healthcare. Archivum Immunologiaeet Therapiae Experimentalis. 2007;**55**:315-327. DOI: 10.1007/s00005-007-0039-1

[57] Kiyama R. Estrogenic terpenes and terpenoids: Pathways, functions and applications. European Journal of Pharmacology. 2017;**815**:405-415. DOI: 10.1016/j.ejphar.2017.09.049

[58] Fan F, Tao N, Jia L, He X. Use of citral incorporated in postharvest wax of citrus

fruit as a botanical fungicide against *Penicilliumdigitatum*. Postharvest Biology and Technology. 2014;**90**:52-55. DOI: 10.1016/j.postharvbio.2013.12.005

[59] Tilaoui M, Ait Mouse H, Jaafari A, Zyad A. Comparative phytochemical analysis of essential oils from different biological parts of *Artemisia herba alba* and their cytotoxic effect on cancer cells. PLoS One. 2015;**10**:e0131799. DOI: 10.1371/journal.pone.0131799

[60] Herrera JM, Zunino MP, Massuh Y, Pizzollito RP, Dambolena JS, et al. Fumigant toxicity from five essential oils rich in ketones against Sitophiluszeamais (Motschulsky). Agriscientia. 2014;**31**:35-41

[61] Manosroi J, Dhumtanom P, Manosroi A. Anti-proliferative activity of essential oil extracted from Thai medicinal plants on KB and P388 cell lines. Cancer Letters. 2006;**235**:114-120. DOI: 10.1016/j.canlet.2005.04.021

[62] Bayala B, Coulibaly AY, Djigma FW, Nagalo BM, Baron S, Figueredo G, et al. Chemical composition, antioxidant, antiinflammatory and antiproliferative activities of the essential oil of Cymbopogonnardus, a plant used in traditional medicine. Biomolecular Concepts. 2020;**11**:86-96. DOI: 10.1515/bmc-2020-0007

[63] Ruben Olmedo VN. Antioxidant activity of fractions from oregano essential oils obtained by molecular distillation. Food Chemistry. 2014;**156**:212-219. DOI: 10.1016/j.foodchem.2014.01.087

[64] Marija M, Lesjak INB. Phytochemical Q20 composition and antioxidant, anti-inflammatory and antimicrobial activities of JuniperusmacrocarpaSibth.et Sm. Journal of Functional Foods. 2014;**7**:257-268. DOI: 10.1016/j.jff.2014.02.003

[65] Privitera G, Luca T, Castorina S, Passanisi R, Ruberto G, Napoli E. Anticancer activity of Salvia officinalis essential oil and its principal constituents against hormone-dependent tumour cells. Asian Pacific Journal of Tropical Biomedicine. 2019;**9**:24-28. DOI: 10.4103/2221-1691.250266

[66] Ju J, Chena X, Xiea Y, Yua H, Guoa Y, Chenga Y, et al. Application of essential oil as a sustained release preparation in food packaging. Trends in Food Science & Technology. 2019;**92**:22-32. DOI: 10.1016/j.tifs.2019.08.005

[67] Luís Â, Ramos A, Domingues F. Pullulan films containing rockrose essential oil for potential food packaging applications. Antibiotics. 2020;**9**:681. DOI: 10.3390/antibiotics9100681

[68] Carpena M, Nuñez-Estevez B, Soria-Lopez A, Garcia-Oliveira P, Prieto MA. Essential oils and their application on active packaging systems: A Review. Resources. 2021;**10**:7. DOI: 10.3390/resources10010007

[69] Ribeiro-Santos R, Andrade M, Sanches-Silva A. Application of encapsulated essential oils as antimicrobial agents in food packaging. Current Opinion in Food Science. 2017;**14**:78-84. DOI: 10.1016/j.cofs.2017.01.012

[70] Wen P, Zhu DH, Wu H, Zong MH, Jing YR, Han SY. Encapsulation of cinnamon essential oil in electrospunnanofibrous film for active food packaging. Food Control. 2016;**59**:366-376. DOI: 10.1016/j.foodcont.2015.06.005

[71] Silva F, Caldera F, Trotta F, Nerín C, Domingues FC. Encapsulation of coriander essential oil in cyclodextrinnanosponges: A new strategy to promote its use in controlled-release active packaging. Innovative

Food Science & Emerging Technologies. 2019;**56**:102177. DOI: 10.1016/j.ifset.2019.102177

[72] Chung SK, Seo JY, Lim JH, Park HH, Yea MJ, Park HJ. Microencapsulation of essential oil for insect repellent in food packaging system. Journal of Food Science. 2013;**78**:E709-E714. DOI: 10.1111/1750-3841.12111

[73] Lee MH, Park HJ. Preparation of halloysite nanotubes coated with Eudragit for a controlled release of thyme essential oil. Journal of Applied Polymer Science. 2015;**132**:42771. DOI: 10.1002/app.42771

[74] Lee MH, Seo HS, Park HJ. Thyme oil encapsulated in halloysite nanotubes for antimicrobial packaging system. Journal of Food Science. 2017;**82**:922-932. DOI: 10.1111/1750-3841.13675

[75] Jang SH, Jang SR, Lee GM, Ryu JH, Park S, Park NH. Halloysite nanocapsules containing thyme essential oil: Preparation, characterization, and application in packaging materials. Journal of Food Science. 2017;**82**:2113-2120. DOI: 10.1111/1750-3841.13835

[76] Biddeci G, Cavallaro G, Di Blasi F, Lazzara G, Massaro M, Milioto S, et al. Halloysite nanotubes loaded with peppermint essential oil as filler for functional biopolymer film. Carbohydrate Polymers. 2016;**152**:548-557. DOI: 10.1016/j.carbpol.2016.07.041

[77] Gorrasi G. Dispersion of halloysite loaded with natural antimicrobials into pectins: Characterization and controlled release analysis. Carbohydrate Polymers. 2015;**127**:47-53. DOI: 10.1016/j.carbpol.2015.03.050

[78] Tornuk F, Sagdic O, Hancer M, Yetim H. Development of LLDPE based active nanocomposite films with nanoclays impregnated with volatile compounds. Food Research International. 2018;**107**:337-345. DOI: 10.1016/j.foodres.2018.02.036

[79] Dong Z, Luo C, Guo Y, Ahmed I, Pavase TR, Lv L, et al. Characterization of new active packaging based on PP/LDPE composite films containing attapulgite loaded with *Allium sativum* essence oil and its application for large yellow croaker (Pseudosciaenacrocea) fillets. Food Packaging and Shelf Life. 2019;**20**:100320. DOI: 10.1016/j.fpsl.2019.100320

[80] Giannakas A. Na-Montmorillonite vs. organically modified montmorillonite as essential oil nanocarriers for melt-extruded low-density poly-ethylene nanocomposite active packaging films with a controllable and long-life antioxidant activity. Nanomaterials (Basel). 2020;**10**:1027. DOI: 10.3390/nano10061027

[81] Cui R, Yan J, Cao J, Qin Y, Yuan M, Li L. Release properties of cinnamaldehyde loaded by montmorillonite in chitosan-based antibacterial food packaging. International Journal of Food Science and Technology. 2021;**56**:3670. DOI: 10.1111/ijfs.14912

[82] Cui R, Zhu B, Yan J, Qin Y, Yuan M, Cheng G, et al. Development of a sodium alginate-based active package with controlled release of cinnamaldehyde loaded on halloysite nanotubes. Foods. 2021;**10**:1150. DOI: 10.3390/foods10061150

[83] Micó-Vicent B, Viqueira V, Ramos M, Luzi F, Dominici F, Torre L, et al. Effect of lemon waste natural dye and essential oil loaded into laminar nanoclays on thermomechanical and color properties of polyester based bionanocomposites. Polymers. 2020;**12**:1451. DOI: 10.3390/polym12071451

[84] Shemesh R, Goldman D, Krepker M, Danin-Poleg Y, Kashi Y, Vaxman A, et al. LDPE/clay/carvacrol nanocomposites with prolonged antimicrobial activity. Journal of Applied Polymer Science. 2014;**132**:41261. DOI: 10.1002/app.41261

[85] Shemesh R, Krepker M, Goldman D, Danin-Poleg Y, Kasi Y, Nitzan N, et al. Antibacterial and antifungal LDPE films for active packaging. Polymers for Advanced Technologies. 2015;**26**:110-116. DOI: 10.1002/pat.3434

[86] Shemesh R, Krepker M, Natan M, Danin-Poleg Y, Banin E, Kashi Y, et al. Novel LDPE/halloysite nanotube films with sustained carvacrol release for broad-spectrum antimicrobial activity. RSC Advances. 2015;**5**:87108-87117. DOI: 10.1039/C5RA16583K

[87] Krepker M, Shemesh R, DaninPoleg Y, Kashi Y, Vaxman A, Segal E. Active food packaging films with synergistic antimicrobial activity. Food Control. 2017;**76**:117-126. DOI: 10.1016/j.foodcont.2017.01.014

[88] Kinninmonth MA, Liauw CM, Verran J, Taylor R, Edwards-Jones V, Shaw D, et al. Investigation into the suitability of layered silicates as adsorption media for essential oils using FTIR and GC–MS. Applied Clay Science. 2013;**83-84**:415-425. DOI: 10.1016/j.clay.2013.07.009

[89] Fernández MA, Roque LB, Espinosa EG, Deyá C, Bellotti N. Organomontmorillonite with biogenic compounds to be applied in antifungal coatings. Applied Clay Science. 2020;**184**:105369. DOI: 10.1016/j.clay.2019.105369

[90] Mohan M, Haider SZ, Andola HC, Purohit VK. Essential oils as green pesticides: For sustainable agriculture. Research Journal of Pharmaceutical, Biological and Chemical Sciences. 2011;**4**:100-106

[91] Oliveira AP, Santana AS, Santana EDR, Lima APS, Faro RRN, Nunes RS, et al. Nanoformulation prototype of the essential oil of *Lippiasidoides* and thymol to population management of *Sitophiluszeamais* (Coleoptera: Curculionidae). Industrial Crops and Products. 2017;**107**:198-205. DOI: 10.1016/j.indcrop.2017.05.046

[92] Varona S, Martín Á, Cocero MJ. Formulation of a natural biocide based on lavandin essential oil byemulsification using modified starches. Chemical Engineering and Processing: Process Intensification. 2009;**48**:1121-1128. DOI: 10.1016/j.cep.2009.03.002

[93] Visavale GL, Nair RK, Kamath SS, Sawant MR, Thorat BN. Granulation and drying of modified clay incorporated pesticide formulation. Drying Technology. 2007;**25**:1369-1376. DOI: 10.1080/07373930701438980

[94] Bhardwaj D, Sharma M, Sharma P, Tomar R. Synthesis and surfactant modification of clinoptilolite and montmorillonite for the removal of nitrate and preparation of slow release nitrogen fertilizer. Journal of Hazardous Materials. 2012;**227-228**:292-300

[95] Shirvani M, Farajollahi E, Bakhtiari S, Ogunseitan OA. Mobility and efficacy of 2,4-D herbicide from slow-release delivery systems based on organo-zeolite and organo-bentonite complexes. Journal of Environmental Science and Health, Part B. 2014;**49**:255-262

[96] Carrizosa MJ, Hermosin MC, Koskinen WC, Cornejo J. Use of organosmectites to reduce leaching losses of acidic herbicides. Soil Science Society of America Journal. 2003;**67**:511-517. DOI: 10.2136/sssaj2003.5110

[97] Nguemtchouin MGM, Ngassoum MB, Chalier P, Kamga R, Ngamo LST, Cretin M. Ocimum gratissimum essential oil and modified montmorillonite clay, a means of controlling insect pests in stored products. Journal of Stored Products Research. 2013;**52**:57-62. DOI: 10.1016/j.jspr.2012.09.006

[98] Nguemtchouin MGM, Ngassoum MB, Kamga R, Cretin M, Chalier P. Insecticidal formulation based on *Ocimumgratissimum* essential oil and montmorillonite clays for maize protection. IOBC-WPRS Bulletin. 2014;**98**:113-121

[99] Nguemtchouin MMG, Ngassoum MB, Ngamo LST, Gaudu X, Cretin M. Insecticidal formulation based on Xylopiaaethiopica essential oil and kaolinite clay for maize protection. Crop Protection. 2021;**29**:985-991. DOI: 10.1016/j.cropro.2010.06.007

[100] NguemtchouinMMG, NgassoumMB, Ngamo LST, Mapongmetsem PM, Sieliechi J, Malaisse F, et al. Adsorption of essential oil components of Xylopiaaethiopica (Annonaceae) by kaolin from Wak, Adamawa province (Cameroon). Applied Clay Science. 2009;**44**:1-6. DOI: 10.1016/j.clay.2008.10.010

[101] Nakhli A, Nguemtchouin MMG, Bergaoui M, Khalfaoui M, Cretin M, Huguet P. Non-linear analysis in estimating model parameters for thymol adsorption onto hydroxyiron-clays. Journal of Molecular Liquids. 2017;**244**:201-210. DOI: 10.1016/j.molliq.2017.08.128

[102] Nguemtchouin MMG, Ngassoum MB, Kamga R, Deabate S, Lagerge S, Gastaldi E, et al. Characterization of inorganic and organic clay modified materials: An approach for adsorption of an insecticidal terpenic compound. Applied Clay Science. 2015;**104**:110-118. DOI: 10.1016/j.clay.2014.11.016

[103] NdomoAF, NgamoLT, TapondjouLA, Tchouanguep FM, Hance T. Insecticidal effects of the powdery formulation based on clay and essential oil from the leaves of *Clausenaanisata* (Willd.) J. D. Hook ex. Benth. (Rutaceae) against Acanthoscelidesobtectus (Say) (Coleoptera: Bruchidae). Journal of Pest Science. 2008;**81**:227. DOI: 10.1007/s10340-008-0211-3

[104] Gueye S, Thiaw C, Sembene M. Insecticidal effect of kaolin powder flavoured with essential oils of *Lantana camara* L. (Verbenaceae) and *Annonasenegalensis Pers*. (Annonaceae) on Caryedonserratus Olivier (Coleoptera-Bruchidae), a groundnut stock pest. International Journal of Biological and Chemical Sciences. 2012;**6**:1792-1797. DOI: 10.4314/ijbcs.v6i4.33

[105] Kéita SM, Vincent C, Schmit JP, Arnason JT, Bélanger A. Efficacy of essential oil of *Ocimumbasilicum* L. and *O. gratissimum* L. applied as an insecticidal fumigant and powder to control *Callosobruchusmaculatus* (Fab.) [Coleoptera: Bruchidae]. Journal of Stored Products Research. 2001;**37**:339-349. DOI: 10.1016/S0022-474X(00)00034-5

[106] Noudem JA, Nguemtchouin MMG, Kaptso KG, Khalfaoui M, Noumi GB. Saponins-clay modified materials: A new approach against *Callosobruchus subinnotatus* in stored products. International Journal of Scientific & Technology & Research. 2017;**6**:134-141

[107] El Miz M, Salhi S, El Bachir A, Wathelet JP, Tahani A. Adsorption of essential oil components of *Lavandula angustifolia* on sodium modified

bentonite from Nador (North-East Morocco). African Journal Of Biotechnology. 2014;**13**:3413-3425. DOI: 10.5897/AJB2013.13450

[108] El Miz M, Salhi S, El Bachir A, Wathelet JP, Tahani A. Adsorption study of thymol on Na-bentonite. Journal of Environmental Solutions. 2013;**2**:31-37

[109] El Miz M, Salhi S, Chraibi I, El Bachiri A, Fauconnier ML, Tahani A. Characterization and adsorption study of thymol on pillared bentonite. Open Journal of Physical Chemistry. 2014;**4**:98-116. DOI: 10.4236/ojpc.2014.43013

[110] Ghrab S, Eloussaief M, Lambert S, Bouaziz S, Benzina M. Adsorption of terpenic compounds onto organo-palygorskite. Environmental Science and Pollution Research. 2018;**25**:18251-11862. DOI: 10.1007/s11356-017-9122-2

[111] Ghrab S, Balme S, Cretin M, Bouaziz S, Benzina M. Adsorption of terpenes from *Eucalyptus globulus* onto modified beidellite. Applied Clay Science. 2018;**156**:169-177. DOI: 10.1016/j.clay.2018.02.002

[112] Gueu S, Tia VE, Bartier D, Barres O, Soro FD. Adsorption of *Lippia multiflora* essential oil on two surfactant-modified clays: Qualitative approach. Clay Minerals. 2020;**55**:219-228. DOI: 10.1180/clm.2020.26

[113] Ziyat H, Bennani MN, Hajjaj H, Qabaqous O, Arhzaf S, Mekdad S, et al. Adsorption of thymol onto natural clays of Morocco: Kinetic and isotherm studies. Journal of Chemistry. 2020;**2020**:1-10. DOI: 10.1155/2020/4926809

[114] Ziyat H, Bennani MN, Allaoui S, Houssaini J, M'Barek HN, Arif S, et al. *In vitro* evaluation of the antifungal activity of ghassoul-based formulations with oregano and thyme essential oils against *Penicillium sp*. Journal of Chemistry. 2021;**2020**:1-8. DOI: 10.1155/2021/6692807

[115] Zhong H, Mu B, Zhang M, Hui A, Kang Y, Wang A. Preparation of effective carvacrol/attapulgite hybrid antibacterial materials by mechanical milling. Journal of Porous Materials. 2020;**27**:843-853. DOI: 10.1007/s10934-020-00863-7

[116] Zhong H, Mu B, Yan P, Jing Y, Hui A, Wang A. A comparative study on surface/interface mechanism and antibacterial properties of different hybrid materials prepared with essential oils active ingredients and palygorskite. Colloids and Surfaces A: Physicochemical and Engineering Aspects. 2021;**618**:126455. DOI: 10.1016/j.colsurfa.2021.126455

[117] Berraaouan D, Elmiz M, Essifi K, Salhi S, Tahani A. Adsorption of carvacrol on modified bentonite in aqueous solutions. Materials Today. 2020;**31**:S28-S32. DOI: 10.1016/j.matpr.2020.05.280

[118] Bernardos A, Bozik M, Alvarez S, Saskova M, Perez-Esteve E, Kloucek P, et al. The efficacy of essential oil components loaded into montmorillonite against *Aspergillusniger* and *Staphylococcus aureus*. Flavour and Fragrance Journal. 2019;**34**:151-162. DOI: 10.1002/ffj.3488

[119] Saucedo-ZuñigaJN,Sánchez-ValdesS, Ramírez-Vargas E, Guillen L, Ramos-deValle LF, Graciano-Verdugo A, et al. Controlled release of essential oils using laminar nanoclay and porous halloysite/essential oil composites in a multilayer film reservoir. Microporous and Mesoporous Materials. 2021;**316**:110882. DOI: 10.1016/j.micromeso.2021.110882

[120] Essifi K, Hammani A, Berraaouan D, El Bachiri A, Fauconnier ML, Tahani A. Montmorillonite nanoclay based

formulation for controlled and selective release of volatile essential oil compounds. Materials Chemistry and Physics. 2022;**277**:125569. DOI: 10.1016/j.matchemphys.2021.125569

[121] Pramanik S, Gopalakrishnan R, Barua N, Buragohain AK, Karak N. Montmorillonite immobilized Curcuma aromatica/Zanthoxylumlimonella oil nanoconjugate as a green antibacterial and biocompatible material with mosquito repellent attributes. Applied Clay Science. 2015;**109-110**:33-38. DOI: 10.1016/j.clay.2015.03.008

[122] Tenci M, Rossia S, Aguzzi C, Carazo E, Sandri G, Bonferoni MC, et al. Carvacrol/clay hybrids loaded into *in situ* gelling films. International Journal of Pharmaceutics. 2017;**531**:676-688. DOI: 10.1016/j.ijpharm.2017.06.024

[123] Nagy K, Bíró G, Berkesi O, Benczédi D, Ouali L, Dékány I. Intercalation of lecithins for preparation of layered nanohybrid materials and adsorption of limonene. Applied Clay Science. 2013;**72**:155-162. DOI: 10.1016/j.clay.2012.11.008

[124] Eusepi P, Marinelli L, Borrego-Sánchez A, García-Villén F, Ben Khalifa R, Cacciatore I, et al. Nano-delivery systems based on carvacrolprodrugs and fibrous clays. Journal of Drug Delivery Science and Technology. 2020;**58**:101815. DOI: 10.1016/j.jddst.2020.101815

[125] Lei H, Wei Q, Wang Q, Su A, Xue M, Liu Q, et al. Characterization of ginger essential oil/palygorskite composite (GEO-PGS) and its anti-bacteria activity. Materials Science & Engineering. C, Materials for Biological Applications. 2017;**73**:381-387. DOI: 10.1016/j.msec.2016.12.093

[126] Calabrese I, Gelardi G, Merli M, Liveri MLT, Sciascia L. Clay-biosurfactant materials as functional drug delivery systems: Slowing down effect in the *in vitro* release of cinnamic acid. Applied Clay Science. 2017;**135**:567-574. DOI: 10.1016/j.clay.2016.10.039

Chapter 3

Organoclay Nano-Adsorbent: Preparation, Characterization and Applications

Kawthar Yahya, Wissem Hamdi and Noureddine Hamdi

Abstract

Organoclay has a tremendous impact on both fundamental studies and practical applications in numerous fields. In this context, this chapter investigates the performance of Organoclay in wastewater treatment. In particular, the adsorption of various hazardous substances has been reviewed. This study aims to give an overview of the preparation methods of Organoclay. The second purpose was to discuss the removal efficiency and reliability of various pollutants by organoclay. The third goal discussed the isotherms and kinetics used for the data interpretation. This work revealed that the characteristics of Organoclay depend mainly on the type of clay used and the nature of the intercalated surfactant. Sorption efficiency was found to depend on the nature of Organoclay, type of pollutant, pH, contact time and the concentration of pollutant.

Keywords: organoclay, adsorption, hazardous substances, wastewater treatment

1. Introduction

Nanotechnology is one of the most prominent promising technologies that offer solutions to different problems in various aspects of our life. In fact, nowadays, nanotechnology used in the synthesis of nanoparticles has attracted great interest in different applications (environment, emerging, industries...etc.) [1–3]. A healthy environment is a major challenge faced by the world today. In connection with the rapid industrialization, scientists reported the presence of more than 700 carcinogenic and highly toxic inorganic and organic micro pollutants. Toxic inorganic metals included instance chromium, mercury, cadmium, lead...etc. and inorganic oxyanions included fluoride, nitrate, phosphate etc. while organic contaminants included, phenols, dye, hydrocarbons, and pesticides. They are considered persistent environmental pollutants non-biotransformable or non-biodegradable [3]. Mining operations, metal plating facilities, textile industries, fertilizer industries and pharmaceutical industries are the most common sources of these hazardous substances [4–8]. Consequently, developing efficient techniques for the treatment of effluent fertilizers and industrial before being discharged into the environment is a considerable issue in terms of public health and environmental protection.

The need for sustainable technologies to conserve the environment leads to the development of a lot of technologies in the field of water treatment [9]. In this scenario, research efforts have been conducted to remove these contaminants from wastewater using several methods such as adsorption, electrolysis electrodialysis, ion exchange, reverse osmosis, coagulation and flocculation, and chemical precipitation have demonstrated different degrees of remediation efficiency [9–11].

Many of these methods, such as coagulation and flocculation, ion-exchange, and reverse osmosis, are expensive and cannot be applied in developing countries. Conventional coagulation methods and chemical precipitation cause secondary contaminants requiring an additional treatment and increasing the treatment cost [3, 11].

Interestingly, adsorption is the most attractive process in developing countries due to its lower cost, easy-to-use and its high efficiency to remove different types of contaminants [3, 12, 13]; thus, the adsorption method produces low quantities of sludge, which is largely produced by other methods (chemical precipitation) [3], but it requires a good choice of adsorbent. Generally, the selection of the useful adsorbent for water treatments is controlled by many factors such as the adsorption capacity of the material toward the target contaminant, cost/efficiency ratio, and the type and concentration of the contaminants present in water [3]. Ideal adsorbents must therefore meet a number of criteria, such as: (1) should be environmentally benign; (2) should demonstrate a high sorption capacity and high selectivity especially to the pollutants occurring in water at low concentration; (3) the adsorbed pollutants can be easily removed from its surface, and (4) should be recyclable [14]. A sorbent with the above characteristics would be considered an excellent adsorbent in wastewater treatment. Efficient adsorbents of biological, organic or mineral origin have been used for wastewater treatment [15, 16]. The most important used adsorbents are agricultural wastes [17], clay minerals [18, 19], modified clays [20, 21], zeolites [22], and activated carbon [23]...etc.

Nanotechnology has great potential for improving the efficiency in preventing water pollution and improving the treatment methods [9]. In fact, the application of nanotechnology in water treatment extends to various fields like nanoscale filtration techniques, adsorption of pollutants on nanoparticles and breakdown of contaminants by nanoparticle catalysis [3, 15].

Among various nanoparticles 'clay minerals' generally speaking, clay minerals which are constituted of layered mineral silicates in the nano dimension, are cheap and non- hazardous, and are characterized by high surface reactivity and stability due to their large surface area [24]. Compared with other adsorbent materials, clay minerals with physical adsorption ability and surface chemical activity, are readily available, making them increasing focus as of late [25, 26]. Nevertheless, the adsorption ability of clay is mainly dependent on fundamental traits, for instance, the charge characteristics of the adsorbent, pH, competing ions and pollutant type [18]. Moreover, clays that exhibit the crystal structure and negative charge have restricted their application [27]. In order to overcome these limitations, many studies in recent years have addressed the improvement of the mechanical properties of clay through the use of various types of modifications [21].

Recently, there has been an increasing interest in using organoclay for the removal of contaminants from soil and aquatic environments due to here the large specific surface and a hydrophobic behavior. Indeed, the intercalation of cationic surfactants changes the surface properties from hydrophilic to hydrophobic, greatly increases the specific surface and increases the basal spacing resulting in exposure of more adsorption sites [28, 29], thus, adsorption capacity especially when surfactant loading exceeds the CEC of clay [30]. Organoclay are a group of surfactant modified clays with hydrophobic properties, which have been extensively used in remediation of heavy metals, herbicides and pesticides, organic compounds and anionic contaminants [31].

The overall aim of this chapter was the investigation of the preparation and characterization of organoclay and their capacities to remove various pollutants as well as empirical findings on the equilibrium isotherms and kinetics. The literature for the chapter included published studies on hazardous substances.

2. Preparation

This section of the chapter describes the preparation of organoclay on the laboratory run, with various experimental conditions, clays from different regions, and various kinds of surfactants. Organoclay material is defined as hybrid materials resulting from clay mineral association with surfactant [28].

Surfactant compounds are amphiphilic molecules, meaning they have two parts of different polarity, one hydrophobic is apolar and the other hydrophilic is polar (**Figure 1**).

Clay minerals are fundamentally hydrous alumino-silicates with highly fine particle sizes. Most clay minerals are made by a stack of sheets; these sheets are made by a stack of tetrahedral and octahedral layers that shape the frame of all clay mineral assemblage; the arrangement of these tetrahedral and octahedral layers makes it possible to discern three main types of clay (1:1 or TO type, 2:1 or TOT type, 2:1:1 or TOTO type) [24]. Clay minerals have different types of physical properties like- the CEC, plasticity, hardness, porosity and adsorption ability (**Figure 1**).

Clay organophilization depends on the characterizations (structure, propriety, type, chemical nature...etc.) of the used clay and surfactant. Among the clay

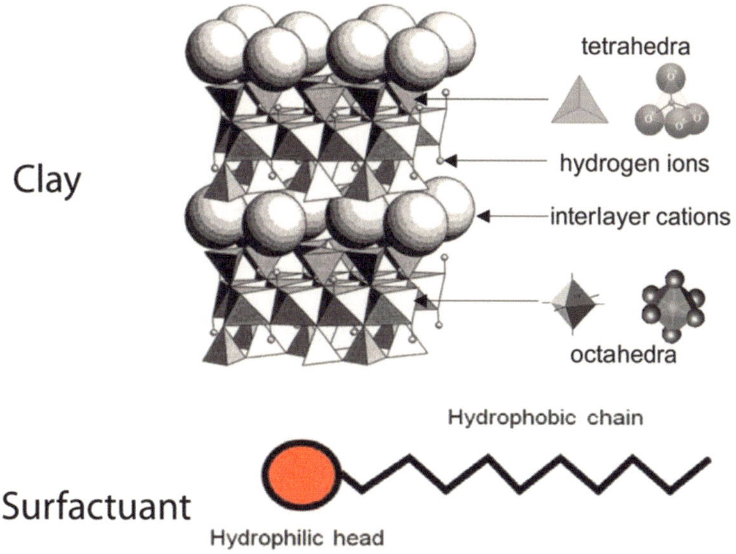

Figure 1.
General structure of a clay mineral [32] and surfactant molecule [33].

minerals, smectites (2:1 or TOT type), have been widely employed to prepare Organoclay since of their excellent properties. For example, several pieces of research [24, 28, 34] are noted among the expandable clay minerals, montmorillonites were the most popular material for the preparation of organoclays because of their singular properties: charge density, cation exchange capacity (CEC) and swelling ability, its abundance in the ground, and thus its low cost. Thus, Jlassi et al. [24] noted a high CEC of clay and a uniform surface charge density promote clay organophilization; therefore montmorillonites are the most favorable swelling clay for the intercalation of organic species in its interlayer space. The compensating cation size (to be replaced) has an impact on clay organophilization. Indeed, the smaller, more mobile, and more easily hydratable the compensating cation are, the easier the exchange is [24]. For this, the purification of the raw clay in order to obtain sodium clay is an essential step in the preparation of organophilic clays. In Fact, Na^+ cations are very exchangeable compared to K^+ and Ca^{2+} cations. The net amount of surfactant adsorbed to the clay minerals can exceed the CEC of the clay minerals. Therefore the entire researcher prepares organoclay with a surfactant concentration exceeding the CEC of the clay used [34].

There are some types of organoclay based on the used surfactant; the best-known surfactants are cationic (quaternary alkylammonium, exotic (zwitterionic) and nonionic ones [28]. For cationic and zwitterionic surfactants, organoclays are obtained by the substitution of the inorganic cations located within the interlayer space through cation exchange, whereas these exchange hydrated cations are kept in the case of nonionic compounds, leading to hybrid materials with a dual hydrophilic hydrophobic behavior [28]. Cationic surfactants were principally used for the preparation of organoclay. The quaternary alkylammonium salts are the cationic surfactants the most preferred for the modification of clay minerals where her arrangements depend on both the length of the alkyl chains and the concentration of the amphiphilic molecules [28]. Therefore, currently, there is a significant several of research on the modification of clay minerals with several kinds of quaternary alkylammonium salts on a laboratory scale.

In literature, various methods are used for the preparation of organoclay such as cation exchange, organosilane, iodonium, and diazonium salt grafting. Of the methods of organophilic preparation, cation exchange is the most commonly employed [24, 34]; cation exchange has been used for five decades. This technique aims to exchange interlayer cations of the clay mineral with surfactant [24, 34].

Experimentally, organoclay can be prepared through three routes in humid states by liquid–solid or liquid–liquid interaction with the use of solvents, or by solid-state reaction with solid–solid interaction without the use of any solvents [24, 34]. The first, a solid–solid process, consists of grinding a mixture of clay powders and surfactant that is subsequently heated to ensure the diffusion of surfactant molecules in the interlayer space [24]. The solid-state reactions of clay minerals and ammonium cations were reported in 1990 [34]. The dry process leads to heterogeneous exchanged organoclays. Even if the intercalation of the compounds (cationic or polar molecules), is confirmed by the expansion of the interlayer space identified by following the 00 l reflection through X-ray diffraction (XRD) [35]. But, the absence of solvents preceding the preparation is environmentally good and makes the process more suitable for industrialization [28]. The second is carried out by a liquid–solid process by putting clay, originally powdered, in contact with an organic cation solution at a known concentration. The third method is a liquid–liquid process and consists of mixing dispersed clay slurry with a solution of organic salt. The easiest way for organoclay preparation, which was reported in many studies, is surely in an aqueous solution.

Indeed, in solution the presence of water surrounding the exchangeable cations amplifies then repulsive forces at long-range order leading to exfoliation of the phyllosilicates sheets offering total access to the entire specific surface area, making easier the adsorption and interaction with surfactants of which chemical nature control the properties of the organoclays [28].

In a general view of the literature, we note that the synthesis is done in two steps, first The preparation starts with a purification step, which is a long and time-consuming process, leading most of the time to sodium clays [24]; which then facilitates the exchange with the molecule of surfactant in the second step (**Figure 2**):

Pre-treatment of clay minerals: The clay fraction (<2 μm) was separated by sedimentation of the clay suspension (based on Stokes Law calculations), and would remove the larger sized particles (>2 μm). Free carbonate remove by attacks with a solution of hydrochloric acid HCl followed by washing several times with water to remove excess HCL; Organic matter was removed by suspension in H_2O_2 followed by washing several times with water to remove excess H_2O_2; finally, treatment with a NaCl solution. This step is essential because firstly it eliminates the impurities (calcite, quartz, etc.) and secondly, it replaces the cations between the sheets with Na + (homoiconic clays). Homoionic clays are made to facilitate exchange; thus avoiding the different degree of exchange and/or competition between the pollutant to be adsorbed and the ions initially present [36].

Preparation of organo-clays: In literature, the most frequently used method is as follows: A given mass in (g) of sodium clay mixes with a given volume in (mL) of a surfactant solution with a concentration multiple of the cation exchange capacity of the sodium clay. This mixture is then stirred for 2 to 24 hours and then centrifuged. The surfactant-modified clay is washed several times with distilled water. The organoclay obtained is dried at 60° C, then ground and sieved (**Figure 2**).

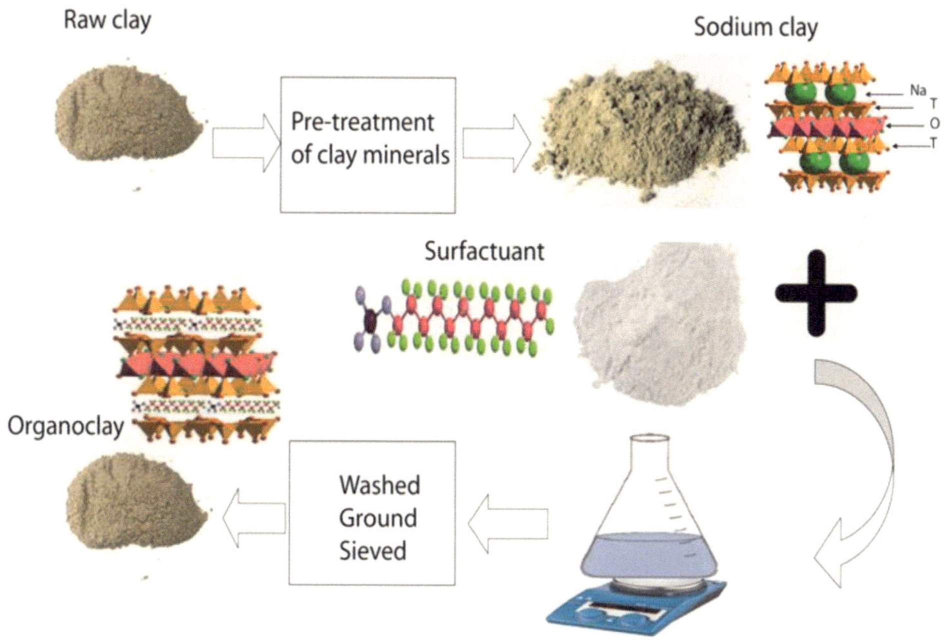

Figure 2.
Preparation of organoclay.

In the literature, most organic clay used in the treatment of water prepares under these conditions: the type of clay used is smectite, the quaternary ammonium ions as surfactant and by an ion-exchange mechanism (Cation exchange); with Surfactant concentration equals 0.5 to 4 times CEC of the clay used.

He et al. [37] used hexadecyltrimethylammonium bromide (HDTMAB) to prepare HDTMA-montmorillonite organoclays to elucidate the relation between the morphology of organoclay and the surfactant packing density within the montmorillonites galleries. The concentrations of HDTMA were 0.5 CEC, 1.5 CEC and 5.0 CEC of montmorillonite, respectively.

Hamdi and Srasra [38] prepared an organo-smectite by cation exchange with hexadecyltrimethylammonium bromide (HDTMAB) at different concentrations (1.0 to 3.0 CEC of clay CEC—cationic exchange capacity). The clay used for preparation was collected from Gabes (southeast of Tunisia).

Msadok et al. [39] modified Tunisian clay with hexadécylpyridinium (HDPy) in a concentration equivalent to 0.5 to 4.0 of cation exchange capacity of purified clay (CEC). This organoclay was also used for application as viscosifier in oil drilling fluid.

Gammoudi et al. [40] studied the influence of exchangeable cation of smectite on surfactant adsorption, Organoclays were prepared Tunisian smectite saturated with Na^{+}, Ca^{2+} and Zn^{2+} ions and the cationic surfactant hexadecyltrimethylammonium bromide.

Yunfei et al. [41] modified sodium montmorillonite by an ion-exchange mechanism using three cationic surfactants: octadecyltrimethylammonium bromide, dimethyldioctadecylammonium bromide, and methyltrioctadecylammonium bromide. These organoclaye were prepared to investigate changes in the surfaces and structures were characterized using X-ray diffraction (XRD), thermal analysis (TG) and infrared (IR) spectroscopy.

Shirzad-Siboni et al. [42] treated montmorillonites with cetyltrimethylammonium bromide (CTAB) by the intercalation method and used it as an adsorbent to uptake herbicide from aqueous solutions.

Guégana and Le Forestierc [43] with the objective of performance evaluation of organoclays for the amoxicillin retention in a dynamic context were modified Na-montmorillonite (Na-Mt) using a set of short and moderate chain surfactant (TMA, BTA, BTMA and TOM), and a long chain organoclay (HDTMA).

De Oliveira and Guégan also prepared organophilic montmorillonite exchanged by various amounts of benzyldimethyltetradecyl ammonium chloride cationic surfactant (BDTAC) up to four times the cation exchange capacity (CEC), to be used for the adsorption of diclofenac [44].

Pandey and De [43] employed a cationic surfactant cetyl tri methyl ammonium bromide (CTAB), wich was prepared in order to explore the adsorption potential of natural bentonite for organic pollutant and anionic molecules.

The solid-state reaction is an alternative route of preparation of organoclays, but it has been less employed than cation exchange. Therefore, **Table 1** summarized 22 important papers in the recent literature, indicating a strong tendency for the use of smectite, quaternary alkylammonium salts and the cation exchange technique to prepare organoclay.

The following sections are organized to study the characterization of these organoclays and their application.

Clay	Surfactant	Concentration	Reference
Smectite	HDPy and HDTMA	0.5 to 3.0CEC	[45]
Bentonite	HDTMA and MTMAB	2CEC	[46]
Montmorillonite	CTAB	1CEC	[42]
Montmorillonites	TMA and HDTMA	40–100% CEC	[47]
Smectite	BDTAC	4.0CEC	[44]
Bentonite	CTAB	—	[43]
bentonite,	HDTMA	2 and 4 CEC	[30]
Bentonite	CTMAB	0.04 to 0.28 CEC	[48]
Bentonite.	CPC	—	[49]
Smectite	HDPy	3.0CEC	[50]
Smectite	HDTMA and (HDPy)	120 and 170% CEC	[51]
Smectite	HDPy	1 to 3 CEC	[52]
Montmorillonites	HDTMA	0.5 to 2CEC	[53]
Montmorillonites	DDAB	—	[54]
Bentonite.	HDPy	—	[55]
Montmorillonites	HDPy	20–400%.	[56]
Montmorillonite	DDTMA and DDDMA)	0.5–2.0 CEC	[57]
Montmorillonite	DTAB and CTAB	1 CEC	[58]
Montmorillonite	BTMA	—	[59]
montmorillonite	HDTMA	20%	[60]
montmorillonite	HDPy	4 CEC	[29]
Bentonite	HDTMA—BTEA	—	[36]

Table 1.
Kinds of clay and surfactant, routes of preparations to prepare organoclay.

3. Characterization

Understanding the structure of organoclays is essential for their practical applications. Consequently, after the synthesis of organoclay, the verification of the intercalation of surfactant in the inter-layer space of the clay by comparison before and after intercalation is essential. In literature, various characterization methods were used on the original clays and the organo-modified clays to get information on structure, expansion capacity, layer charge, pore size, crystallite size, charge distribution and pore distribution. The indispensable methods for characterization are X-ray diffraction (XRD), Fourier transforms infrared spectroscopy (FTIR), Scanning electron micrographs (SEM) and transmission electron microscopy (TEM) and determination of physic-chemical properties (CEC, area-specific, pHpzc ... etc.).

3.1 X-ray diffraction (XRD)

Intercalation of organic surfactant between layers of clays greatly changes (increases) the basal spacing of the layers; X-ray diffraction was used to study this

Changement. Otherwise, X-ray diffraction (XRD) can give the basal spacing (d_{001}) information of the organo-clays which is very important for explaining the intercalation and configuration of surfactant between clay layers. The arrangement of cationic surfactant in the interlayer space of clay minerals was initially deduced in 1969 [34]. Generally speaking, on the basis of XRD results for raw clay the into-foliar cations only form monolayers; but in the case of organoclay, the organic cations (surfactant) may form monolayers, bilayers and paraffin-type layers. The arrangement of organic cations (surfactant) in organoclay depends on the layer charge (=interlayer cation density = packing density of the alkylammonium ions) of the clay mineral and the chain length of the organic ion. On the other hand, with an increasing concentration of added surfactants, the arrangement of surfactant change from monolayer to paraffin-type layers (**Figure 3**) [24, 28, 34].

For example, Msodok et al. [39] notated that according to the literature, the thickness of the montmorillonite is 9.7 Å and the molecular size of HDPy is approximately 23.1 Å in length and 4.6 Å in height. At a small concentration of the cationic surfactant, the interlayer spacing (d_{001}) of 0.5 CEC was 14.4 Å which is attributed to monolayer arrangement. For the 1.0 CEC sample, the basal spacing was 21.96 Å which is assigned to the pseudo-trimolecular arrangement. From 2.0CEC, 3.0CEC and 4.0 CEC data, the basal spacing increased in respect of 0.5 CEC-A with the maximum (d001) was 44.51 Å for 4.0 CEC. These data indicate that the surfactant molecules are located as paraffinic bilayer arrangement in the interlayer space of the montmorillonites (**Figure 4**) [39]. This result is in accordance with another study [45, 61].

Also, He et al. [37] and Hamdi and Srasra [38] prepared an organoclay by cation exchange with hexadecyltrimethylammonium bromide at different concentrations. The X-ray diffraction analysis of various organoclays prepared indicates the basal spacings are expanded as expected depending on the surfactant concentrations. After

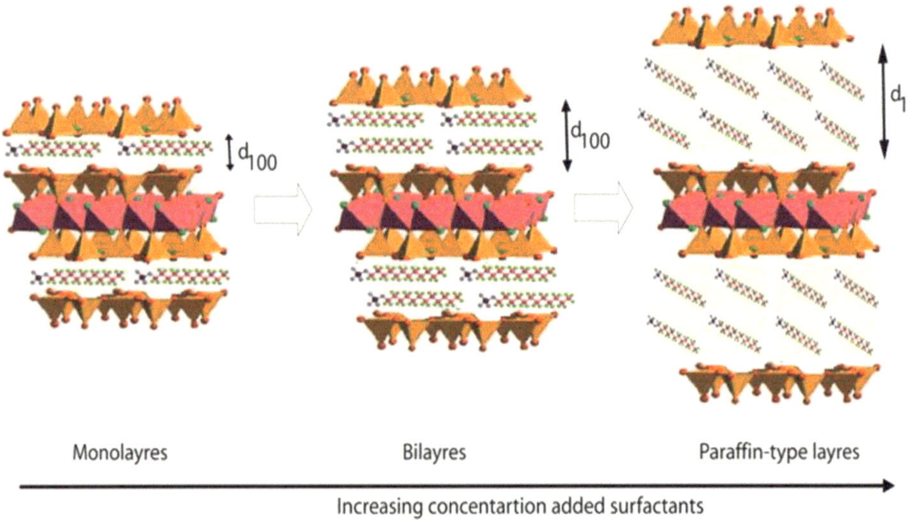

Figure 3.
Arrangement of surfactants in the interlayer space of a clay.

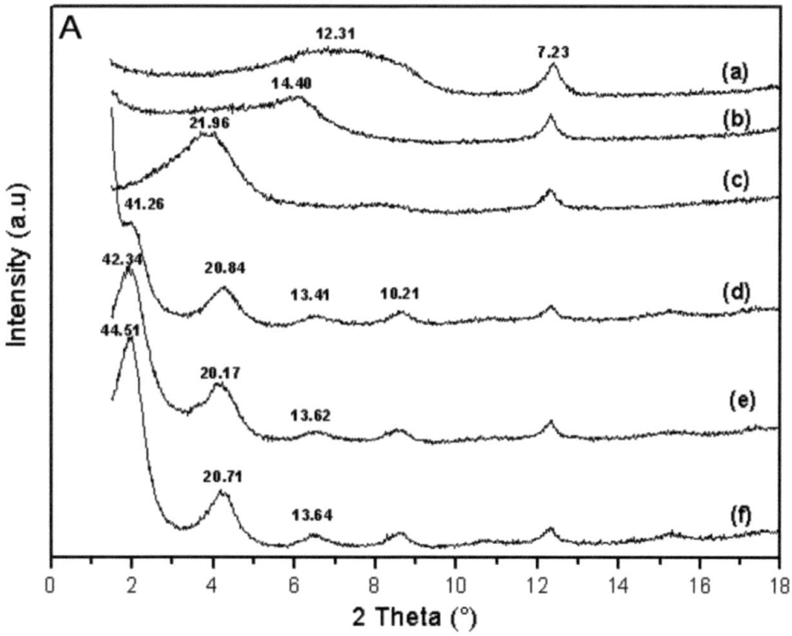

(a) raw clay, (b) 0.5 CEC (c) 1.0 CEC ,(d) 2.0 CEC, (e) 3.0 CEC, (f) 4.0 CEC

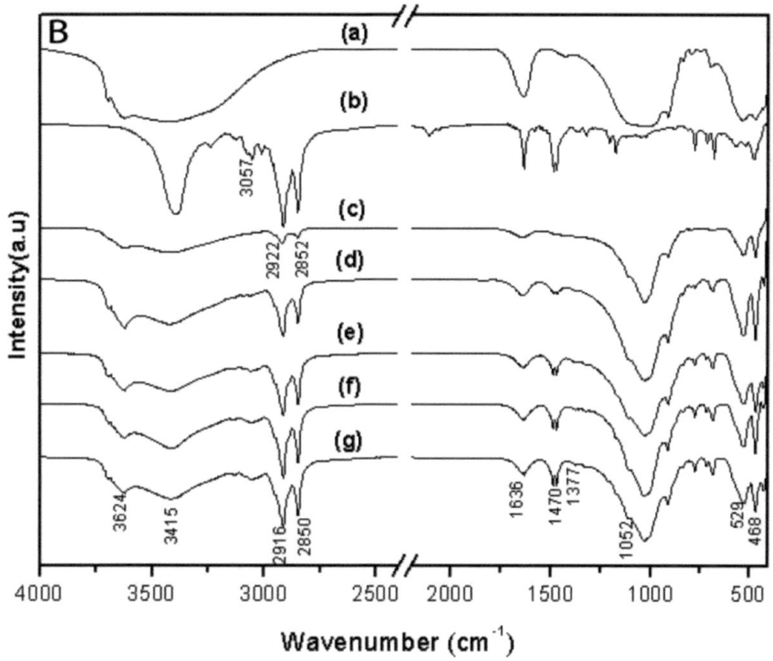

(a) raw clay, (b) HDPy, (c) 0.5 CEC , (d) 1.0 CEC, (e) 2.0 CEC, (f) 3.0 CEC, (g) 4.0 CEC

Figure 4.
(A) XRD patterns; and (B) FTIR spectra of clay and organoclays [39].

treatment with a surfactant (0.5, 1 and 2 CEC of clay) the peak d001 of smectite at 12.69 Å passed to 17.59, 19.25 and 21.62 Å, respectively. The increase in the basal spacing of sodium clay with HDTMA cations can be attributed to the replacement of the inorganic interlayer cations and their hydration water with HDTMA cations [38].

Xi et al. prepared Organoclays based on halloysite, kaolinite and bentonite and used DRX and IR for characterization. XRD pattern has shown that the exchange of cations by surfactant causes the expansion of different clay layers. The expansion is proportional to the concentrations of HDTMA used and varied between the types of clay. For bentonite modified with 2CEC one dominant reflection at 20.2 Å was noted corresponding to basal spacing with surfactant molecules laying flat between clay mineral layers. But For bentonite modified with 4 CEC two d_{001} spacings were observed at18.5 Å corresponding to the bentonite expanded with HDTMA molecule laying flat between two layers and 35.7 Å attributed to the surfactant molecules at right angles to the clay mineral surface. On the other hand, the HDTMA modified kaolinite and halloysite did not show much change in XRD patterns compared to those of untreated ones [30].

3.2 Fourier transforms infrared spectroscopy (FTIR)

Infrared spectroscopy is a powerful tool for the study of the bonding mechanisms on a molecular scale. In FTIR the cationic surfactant is characterized by the presence of C–N (vibrations in tertiary amines) and C\H stretching bands and aliphatic C\H stretching bands of CH_2. In fact, Fourier transforms infrared spectroscopy (FTIR) will further confirm the presence of organics in the clay materials and offers the additional advantage of confirming the configuration of the organic molecule.

Generally, if examining the spectra of raw clay compared to the organoclay; noting the presence of similar bands between both clay such as (H–O–H stretching, H–O–H bending, stretching mode of Si–O...etc.). Thus, additional bands in organoclay: C–N vibrations of tertiary amines and of C–H stretching bands and aliphatic C–H of CH_2. The intensity and width of the similar and additionally bands of absorbance for raw clay and organoclay exhibit distinct differences. Generally, the decrease of band intensity of the OH stretching and bending vibrations are explained by the replacement of the hydrated cations with cationic surfactants [62], which indicates the change of surface clay to the organophilic character. Thus, the increase of surfactant concentration added in clay engenders a slight shift for the symmetric bands of CH_2. Generally, this shift of CH_2 antisymmetric frequency was used to identify the environment surfactant in the interlayer space of organoclay [45]. The higher frequencies indicate a liquid-like environment while the lower frequencies indicate a solid-like environment.

Using infrared spectroscopy, Gammoudi et al. [45] compared the spectrum of raw clay and organic clay showing the presence of similar bands and new bands in organic clay-like as the peaks at 1463 and 1473 cm^{-1} indicate the presence of C–N vibrations in tertiary amines and the N–H stretching peak at 3016 cm^{-1} appeared only after the addition of HDTMA at concentrations greater than 3 CEC and the peak of C\H stretching bands of CH_2 and aliphatic C\H stretching bands. The frequency and the intensity of asymmetric and symmetric stretching bands of CH_2 change with the amount of intercalated surfactant. This indicates that, as the loading surfactant on sodium clay increased, the confined amine chains changed to gauche conformation for trans conformation (to lateral arrangement for paraffin arrangement). This finding is in concordance with previous studies [39] (**Figure 4**).

Thus in the work of Shirzad-Siboni et al. [42], the comparison of FTIR spectra of organoclay with raw clay exhibits significant changes in some of the peaks. In particular, the shift in the siloxane peak after loading of surfactant, the additional peaks appointed to –CH–stretching vibration, could be observed only in organoclay. Additionally are noted that the band at 3460 cm^{-1} disappeared after the modification of MMT nonmaterial with CTAB, which indicates the removal of water molecules and the change in the hydrophobicity of MMT nonmaterial. Also, is noted that with the loading of surfactant asymmetric (CH_2) shifts from 2927 to 2922 cm^{-1} and symmetric (CH_2) shifts slightly from 2856 to 2852 cm^{-1} for organoclay. This indicates that, in the existence of added surfactant, the confined surfactant chains adopt a fundamentally all-trans conformation. This result clearly reveals that the surface modification of MMT is achieved by surfactants.

3.3 Scanning electron micrographs and transmission electron microscopy

SEM and TEM are used to collect detailed information on the morphology.

Msodek et al. [39] used SEM to determine the change in surface morphology of raw clay upon the intercalation of HDPy surfactant. Results showed that the photomicrography of raw clay exhibits massive and aggregate morphology (**Figure 5**). While, at low surfactant concentration (1.0 CEC $\leq$), the particles exhibit a compact form that can be explained by the interactions between the R groups of alkyl chains of surfactant. For the concentrations in exceeded the CEC of clay, the particles were changed to flat layers.

Using SEM and TEM, Pandey and De observed that the raw clay (bentonite) showed rough surface morphology however organo-bentonite showed a smooth surface with large size particles.

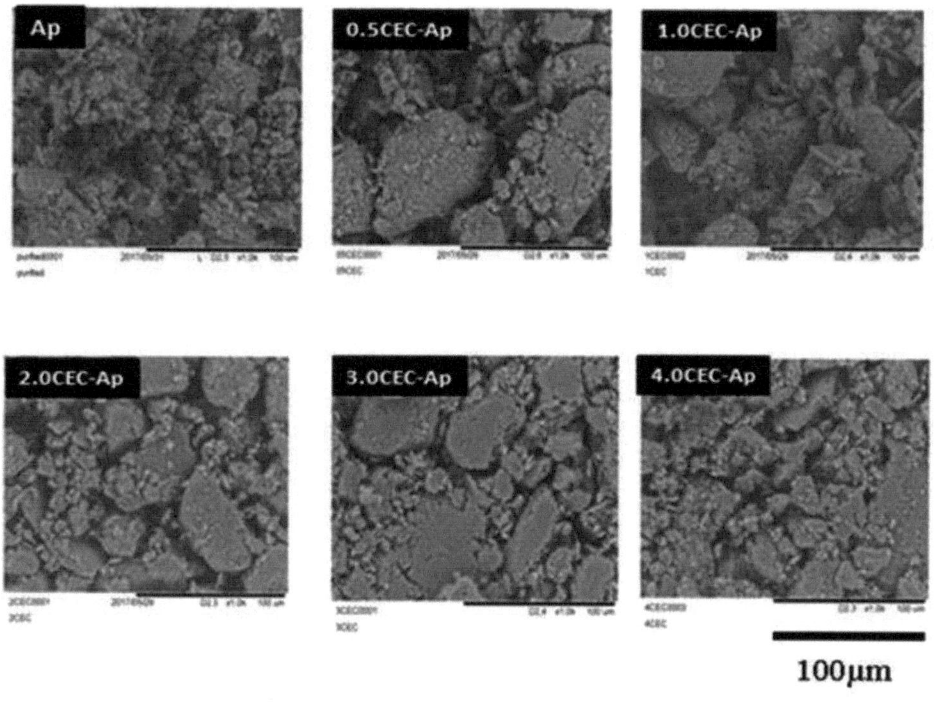

Figure 5.
Photomicrographs of clay and organoclays [39].

Adsorbent	SBET	Vp	CEC	pH_{PZC}	Pore size	Reference
Organosmectites						
0.5CEC	14.83	0.0861	56	7	—	[38]
1CEC	3.42	0.0157	33	7.7	—	
2CEC	1.96	0.0006	28	8	—	
Bentonite	33.02	0.10	97.6	—	11.52	[43]
CTAB–bentonite	10.40	0.04	—	—	16.11	
Na- montmorillonites	29.089	0.035	—		—	[64]
DTAB- montmorillonites	13.345	0.028	—	6.5	—	
CTAB- montmorillonites	6.036	0.022	—	11.2	—	
Na-Bentonite	36.5		0.52	4.5		[65]
Organoclay	27.9		0.72	7.8		

Table 2.
Physicochemical properties of the raw clay and organoclay.

Thus Shirzad-Siboni et al. [42] employed SEM analysis to evaluate the surface morphology of the raw clay (montmorillonites,MMT) and organoclay (montmorillonite). Results reveal that the surface morphologies of both raw clay and organoclay have an uneven structure with non-uniform size distribution. The MMT shows massive, aggregated morphology, while, after modification, the clay surface was changed to a non-aggregated morphology with severely crumpled structures. Furthermore, the surface of Organoclay was expanded.

3.4 Physicochemical properties

The intercalation of surfactants in clay changes their Physicochemical properties (CEC, pore volume, Pore size, pH_{PZC}...etc.).

Hamdi et al. noted that specific surface area decreases significantly as a function of the increase of exchanged surfactant in the specimen. The same results were obtained for the CEC and The pore volume (Vp). However, the value of PZC increased with the increase of exchanged surfactant in the raw clay. The SSA decrease indicates a compact packing of larger surfactant molecules between the silicate layers of mineral clay and blocking the passage of nitrogen molecules [63]. The same outcomes were reported by Pandey and De noted that The Pore size, surface area, as well as micro and mesopore volume, was drastically reduced after cationic surfactant treatment (**Table 2**).

4. Applications

Organoclays have been used in various applications. These applications include adsorbents, rheological control agents, paints, grease, cosmetics, personal care products, and oil well drilling fluids. Currently, an important application of the organoclays is in pollutant adsorption. In fact, a view of the literature shows that various studies have shown that replacing the inorganic exchange cations of clay minerals with organic cations can result in greatly enhanced abilities to remove various contaminants from water:

4.1 Application of organoclay to remove fluoride, phosphate and nitrate

Phosphate, fluoride and nitrate are essential nutrients in the aquatic environment, but excessive phosphate (above 0.02 mg/L), nitrate (above 0.02 mg/L) and Fluoride (above 1.5 mg/L) input may lead to eutrophication (causing degradation of water quality) respectively. Nowadays, the presence of fluoride, phosphate and nitrate in the environment has been identified as one of the acute problems worldwide. Organoclay has been successfully utilized for the removal of nitrate, phosphate and fluoride:

Gammoudi et al. prepared organoclay with smectite using two cationic surfactants (HDPy and HDTMA) for fluoride removal from wastewaters. The optimal condition suitable for defluoridation consists of low fluoride concentration, contact time equal to 6 h and acidic pH. These organoclays showed high removal capacities for fluoride ions [45].

Xi et al. prepared various Organoclays with on halloysite, kaolinite and bentonite for removal of nitrate. The results showed that all these raw clays showed poor adsorption amounts for nitrate. However, when these clays were modified with surfactant HDTMA in 2 or 4 CEC, the removal amounts of these clays were greatly improved. Thus, among all these organoclays, HDTMA modified bentonite showed the best result [30].

Ma and Zhu [48] used inorganic–organo-bentonite which is prepared by replacing the exchangeable inorganic cations with cetyltrimethylammonium bromide (CTMAB) to phosphures from water. The results of this study reveal that inorganic–organo-bentonite can act as a successful adsorbent for removing the phosphures from contaminated water. More than 95% phosphate and 99% phenanthrene were removed from the water within 30 min. The amount of sorbed phosphate increased as pH decreased and the sorption amount increased slightly with an increase in temperature [48].

4.2 Application of organoclay to remove dye

Dyes and pigments are widely used by several industries like plastics, textile, cosmetics...etc. Textile dyeing process is an important source of contamination responsible for the continuous pollution of the environment. In recent years Organoclays have been attractive for use as selective sorbents for dyes:

Tabak et al. [49] studied the adsorption of the Reactive Red 120 by cetylpyridinium modified Resadiye bentonite (CP-bentonite) prepared by ion exchange at different temperature, force ionique and pH levels. Their results showed that the structural arrangement of cetylpyridinium ions in the CP-bentonite sample, as well as the pH, temperature and ionic strength of the bulk solution, influenced the adsorption of RR 120 dye from aqueous solutions by CP-bentonite [49].

Ma et al. [46] showed that bentonite modified with hexadecyltrimethylammonium bromide could be used as a highly efficient adsorbent for the removal of acid dyes from an aqueous solution. The study reveals that the adsorption capacity of organobentonites is affected by the surfactant alkyl chain length [46].

Gammoudi and Srassra [50] studied the removal of three dyes (methyl orange (MO), indigo carmine (IC) and phenol red (PR) on the surface of Tunisian

HDPy-modified; the maximum adsorption capacity from the Langmuir equation was calculated at 227.27, 326.40, and 344.82 mg/g, respectively. The results revealed that the kinetics uptake was fast and equilibrium was attained within 30 min for IC, PR and 60 min for MO.

4.3 Application of organoclay to remove pesticides and herbicide

Shirzad-Siboni et al. [42] were prepare surfactant-modified pillared montmorillonites using cetyltrimethylammonium bromide and used them as adsorbents to remove bentazon from aqueous solutions. Results showed that the maximum adsorption capacity was estimated to be 500 mg/g at pH 3 and room temperature. The removal efficiency at optimum pH 3 was found to increase with the increase in contact time and adsorption dosage, but to decrease with an increase in initial bentazon concentration [42].

Carrizosa et al. [41] examined the adsorption of the herbicide dicamba by organoclays at various concentrations and pH levels. Their results showed that the adsorption capacity of organoclays is favored for high-layer charge and saturation with bulky organic cations close to the CEC.

4.4 Application of organoclay to remove heavy metal

Some heavy metals are notorious water pollutants with high toxicity and carcinogenicity. Heavy metals have been removed from water by adsorption technology using nanoparticles. Organoclay is effective for the removal of various heavy metals. The elimination efficacies of these contaminants are discussed below.

HDTMA modified natural kaolin has been confirmed effective sorbent for Cr (VI) with a maximum adsorption capacity of 27.8 mg/g, while the unmodified natural kaolin was only 0.7 mg/g [60]. Results indicate that most of the adsorption of chromate occurred through the anion exchange of the Br-anion of the HDTMA ion pair [66].

Similarly, the adsorption scheme of oxygen anions of Cr and Mowas studied with bentonite modified by cetylpyridine bromide (CPBr). The equilibrium adsorption capacity of Mo (VI) of 1.4 mmol/g quantity to double the capacity of Cr(VI) (0.7 mmol/g), indicating that Mo (VI) was removed on organic bentonite in the form of polynuclear anions [67].

4.5 Application of organoclay to remove pharmaceutical contaminants

Various researches have reported the successful elimination of pharmaceutical pollutants using organoclay:

Saitoh and Shibayama synthesized organo-clays: Didodecyldimethylammonium bromide (DDAB)-montmorillonite which was used for the removal of 1β-lactam antibiotics from water. The study showed that the removal of antibiotics increased with increasing the amount of organoclay added and the amount of DDAB sorbed on MT. Thus, the authors postulated that the Didodecyldimethylammonium bromide (DDAB)-montmorillonite (MT) organoclay was a useful sorbent not only for the removal of β-lactam antibiotics from water but for their eco-friendly degradation [54].

Polubesova et al. [68] examined tetracycline and sulfonamide antibiotics sorption by the BDMHDA modified montmorillonite. Their results indicated that BDMHDA modified montmorillonite is very efficient for water purification from tetracycline and sulfonamide antibiotics [68].

Guégan and Le Forestier [47] used modified montmorillonite with tetramethyl ammonium (TMA) and hexadecyl trimethyl ammonium (HDTMA) cationic surfactants as adsorbents for the retention of amoxicillin (AMX) [47].

4.6 Application of organoclay to remove hydrocarbons

Masooleh et al. [53] studied the performance of a organically modified nanoclay for petroleum hydrocarbon adsorption. The obtained results that the adsorption capacity of the organoclay was clearly higher than that of the unmodified clay and the hydrocarbons the adsorption capacity was in the range of 4 to 10 g of adsorbent. Also, adsorption equilibrium was attained within 1 h.

4.7 Application of organoclay to remove phenol

Park et al. [57] prepared two types of organoclays from different surfactants (DDTMA and DDDMA) for the adsorption of phenolic compounds. This study revealed the potential utility of the organoclays as adsorbents for the uptake of industrial pollutants in environmental applications.

Zhang et al. [64] were prepared organoclays using withdodecyltrimethylammonium bromide (DTAB) and cetyltrimethylammonium bromide (CTAB); were used as adsorbents for 4-chlorophenol and 2,4-dichlorophenol. This study demonstrates that the adsorption process is affected by the initial solution pH and temperature.

4.8 Application of organoclay to remove radioactive

Bentonite clay is suggested as a buffer material in various concepts for repositories for high-level radioactive waste. Two different mechanisms have been established to study the importance of the ionic balance between the interlayer space of montmorillonites and an external solution: the Donnan equilibrium and the ion exchange equilibrium [69].

Yang et al. [55] synthesized Hexadecylpyridinium Chloride Monohydrate modified bentonite (HDPy-bent) and used it as an adsorbent to remove-99 (99Tc). Results are demonstrated that the HDPy-bent is a low-cost sorbent which can efficiently eliminate Technetium-99 from wastewaters.

Li et al. [58] examined the efficacy of organoclay as sorbents to bind iodide (I^-) and iodate (IO_3^-) from groundwater. Results showed that these sorbents were highly effective at removing I^- and IO_3 from groundwater under oxic conditions, with the adsorption capacity up to 30 mg I/g sorbent.

Some studies, evaluate the efficiency and capability of organoclay to simultaneously remove various pollutants from wastewater:

Yahya et al. [61] used smectitic clay modified using cationic surfactant (hexadecylpyridinium for the removal of fluoride and phosphate single and industrial aqueous solutions. The results show low adsorption ability for fluoride and phosphate ions comparing the case of single solutions. Thus, the selectivity of fluoride is better than phosphate and with coexisting chloride, sulfate ions and other cations.

Xiaoying et al. [36] are studying the possibility of organoclays used to simultaneously remove amoxicillin (AMX) and Cu (II) from wastewater. Results showed that the adsorption of AMX onto organ-bentonite was 6 times higher than that using

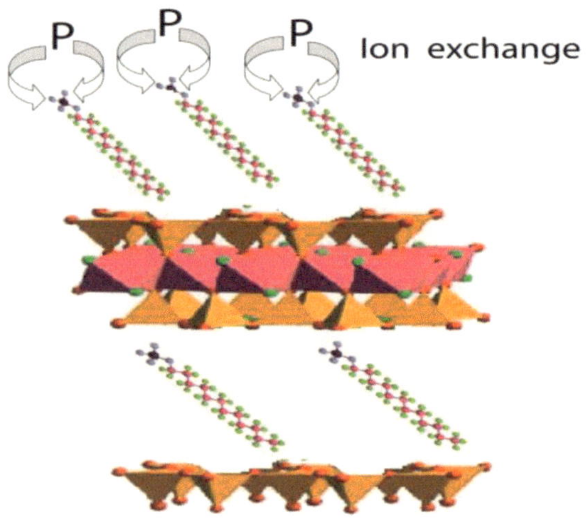

Figure 6.
Mechanism of pollutant (P) uptake in organoclay.

bentonite (control), while adsorption of Cu (II) on organ-bentonite showed comparable results as that using bentonite. The simultaneous adsorption of AMX and Cu (II) onto organ-bentonite occurred through partition for AMX and ion-exchange for Cu (II). More than 34.8% AMX and 43.6% Cu (II) were removed from industrial wastewater, indicating its great potential removal of mixed organic and metal contaminants.

Oyanedel-Craver et al. [36] were studying the feasibility of using hexadecyltrimethylammonium bentonite clay (HDTMA-clay) and benzyl triethylammonium bentonite clay (BTEA-clay) for simultaneous sorption of benzene and one of four heavy metals (Pb, Cd, Zn and Hg). Results showed that both organoclays tested had dual sorptive properties for both heavy metals and an organic contaminant. But, sorption of Pb, Cd, and Zn on both BTEA- and HDTMA-clay decreases in the presence of benzene relative to the sorption obtained without benzene present.

Generally, the predominant mechanism of the adsorption of dye, inorganic oxyanions (fluoride, phosphate nitrate) is an ion exchange [46, 51, 61] (**Figure 6**). The mechanisms mainly include electrostatic adsorption of heavy metal, redox, ion exchange, precipitation, coordination (chelation), as well as surface complexation. For heavy metals, the mechanisms mainly include surface complexation, redox, ion exchange, precipitation, coordination (chelation) and electrostatic adsorption (**Figure** 7) [26].

Table 3 represents various applications of organoclay for the removal of wastewaters pollutants by adsorption.

5. Conclusion

The study of organoclays is a vast field and shows immense potential to be explored. This chapter describes the various ways of preparation of organoclays with the use of cationic surfactants (quaternary alkyl ammonium). Generally, organoclay is obtained by the replacement of the inorganic cations through cation exchange with surfactant. Organoclay can be made using a variety of clay minerals; however,

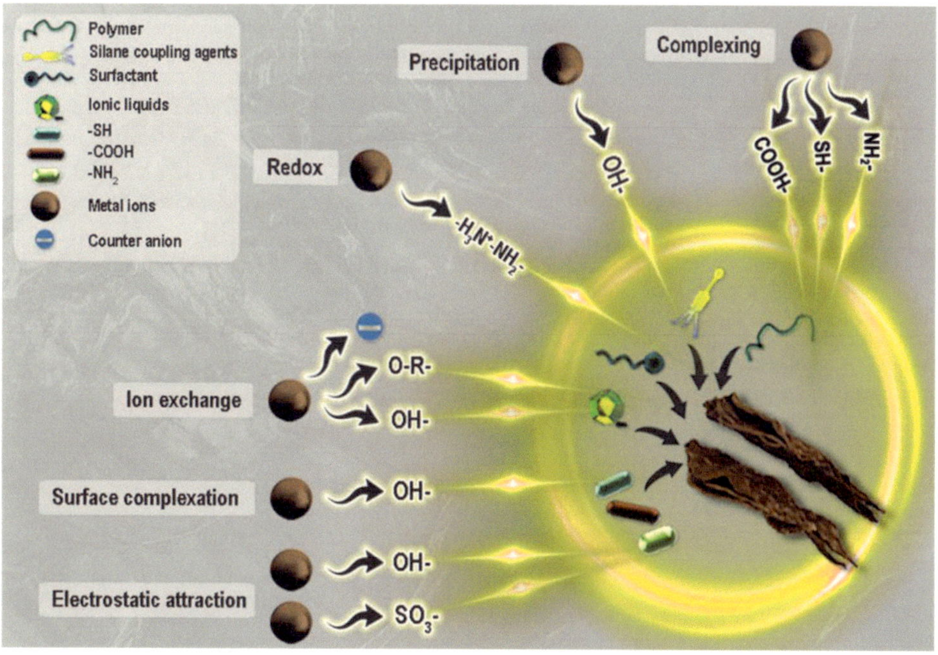

Figure 7.
Adsorption mechanisms of functional organoclays for heavy metal ions [26].

smectite is the most commonly used due to its specific properties. Thus, quaternary ammonium cations are most commonly used as surfactants.

A detailed understanding of the structure of organoclay is of importance in its design and applications. In this chapter, FTIR, transmission electron microscopy (TEM), scanning electron microscopy (SEM) and X-ray diffraction (XRD) has been employed to provide insights into the interlayer structure and morphology of organoclay. The details are as follows:

- XRD patterns show that the interlayer spacing increased with the increase of the added surfactant concentration.

- FTIR spectra confirm these results were by the changes in frequencies and intensities of symmetric and antisymmetric stretching bands of $-CH_2$.

- The intercalation of surfactants in clay decreases the specific surface area, CEC and pore volume. However, increases the value of pH ZPC.

Using SEM and TEM indicate that the raw clay showed rough surface morphology however organoclay showed a smooth surface with large size particles.

Organoclays can be used in various applications including adsorbent systems in the environmental field. These adsorbents display interesting adsorption properties for several organic compounds, especially hydrophobic chemicals.

Despite the potential interest in environmental applications, the use of organoclays appears to be limited to batch experiments but has not yet been explored under dynamic conditions that reduce the efficiency of adsorption.

Pollutants	Removing capacities	Isotherm study	Kinetics study	Mechanism	Reference
Dye	144.08–249.62	—	—	ion-exchange	[46]
Dye	227.27–344.82	Langmuir	pseudo-second order	ion-exchange	[50]
Dye	81.97	Langmuir	pseudo-second order	ion-exchange	[49]
Dye	324.36–399.74	Freundlich	pseudo-second order	—	[65]
Heavy metal	11.970	Langmuir	pseudo-second order	Complexation	[70]
Heavy metal	—	Freundlich	pseudo-second order	ion-exchange	[71]
Herbicide	500 mg/g	Langmuir	pseudo-second order	—	[42]
Nitrate	(0.287 meq/g	Langmuir	pseudo-second order	—	[72]
Chromate-molybdate	0.7–1.4 mmol/g	—	pseudo-second order	ion-exchange	[73]
chlorophenols	458.2–585.80	Langmuir	pseudo-second order	Complexation	[64]
Phenol	—	Freundlich	pseudo-second order	various mechanisms	[57]
Iodide and iodate	21.1–27.	Langmuir	—	—	[58]
Antibiotics	96–99.9%	Langmuir	—	—	[68]
Fluoride	14.02	Langmuir	pseudo-second order	ion-exchange	[45]
Nitrate	2.63–15.38	Langmuir	pseudo-second order	ion-exchange	[74]
Dey	49.53 mg g^{-1}	—	—	—	[43]
Nitrate-perchlorate	0.6–1.1 $mmolg^{-1}$	Langmuir	pseudo-second order	—	[29]
Phosphate-fluoride	11.36–27.77	Langmuir	pseudo-second order	ion-exchange	[61]
Amoxicillin-Cu(II)	34.8–43.6%	Langmuir Freundlich	pseudo-second order pseudo-first order	partition ion-exchange	[60]

Table 3.
Applications of organoclay for the elimination of water pollutants by adsorption.

Finley, overall, in this chapter researchers suggest that the Organoclay could be considered a cheap and efficient adsorbent for the removal of most of the chemical pollutants from wastewater that could be of socioeconomic and environmental relevance.

Author details

Kawthar Yahya[1,2]*, Wissem Hamdi[3] and Noureddine Hamdi[1,2]

1 Higher Institute of Water Sciences and Techniques of Gabes, University of Gabes, Zrig Eddakhlania, Tunisia

2 National Center of Research in Materials Sciences, Borj Cedria Technopole (CNRSM), Soliman, Tunisia

3 High Institute of Agronomy, Sousse University, Sousse, Tunisia

*Address all correspondence to: kawther.y@yahoo.com

References

[1] Osagie C, Othmani A, Ghosh S, Malloum A, Kashitarash Esfahani Z, Ahmadi S. Dyes adsorption from aqueous media through the nanotechnology: A review. Journal of Materials Research and Technology. 2021;**14**:2195-2218. DOI: 10.1016/j.jmrt.2021.07.085

[2] Ogunfowora LA, Iwuozor KO, Ighalo JO, Igwegbe CA. Trends in the treatment of aquaculture effluents using nanotechnology. Cleaner Materials. 2021;**2**:100024. DOI: 10.1016/j.clema.2021.100024

[3] Amari A, Mohammed Alzahrani F, Mohammedsaleh Katubi K, Salem Alsaiari N, Tahoon MA, Rebah FB. Clay-polymer nanocomposites: Preparations and utilization for pollutants removal. Materials. 2021;**14**(6):1365. DOI: 10.3390/ma14061365

[4] Dutta R, Nagarjuna TV, Mandavgane SA, Ekhe JD. Ultrafast removal of cationic dye using Agrowaste-derived mesoporous adsorbent. Industrial and Engineering Chemistry Research. 2014;**53**(48):18558-18567. DOI: 10.1021/ie5030003

[5] Lin J, Zhan Y, Zhu Z. Evaluation of sediment capping with active barrier systems (ABS) using calcite/zeolite mixtures to simultaneously manage phosphorus and ammonium release. Science of the Total Environment. 2011;**409**(3):638-646. DOI: 10.1016/j.scitotenv.2010.10.031

[6] Neşe Ö, Ennil KT. A kinetic study of nitrite adsorption onto sepiolite and powdered activated carbon. Desalination. 2008;**223**(1-3):174-179. DOI: 10.1016/j.desal.2007.01.209

[7] Sari A, Tuzen M. Cd(II) adsorption from aqueous solution by raw and modified kaolinite. Applied Clay Science. 2014;**88-89**:63-72. DOI: 10.1016/j.clay.2013.12.021

[8] Tamtam F et al. Assessing the fate of antibiotic contaminants in metal contaminated soils four years after cessation of long-term waste water irrigation. Science of the Total Environment. 2011;**409**(3):540-547. DOI: 10.1016/j.scitotenv.2010.10.033

[9] Tom AP. Nanotechnology for sustainable water treatment—A review. Materials Today: Proceedings. 2021:S2214785321042565. DOI: 10.1016/j.matpr.2021.05.629

[10] A. Elena et al., « Waste water treatment methods », in Water Treatment, W. Elshorbagy, Éd. London: InTech, 2013. doi: 10.5772/53755.

[11] Zajda M, Aleksander-Kwaterczak U. Wastewater treatment methods for effluents from the confectionery industry—An overview. Journal of Ecological Engineering. 2019;**20**(9):293-304. DOI: 10.12911/22998993/112557

[12] Zhu R, Chen Q, Zhou Q, Xi Y, Zhu J, He H. Adsorbents based on montmorillonite for contaminant removal from water: A review. Applied Clay Science. 2016;**123**:239-258. DOI: 10.1016/j.clay.2015.12.024

[13] Singh NB, Nagpal G, Agrawal S, Rachna. Water purification by using adsorbents: A review. Environmental Technology & Innovation. 2018;**11**:187-240. DOI: 10.1016/j.eti.2018.05.006

[14] Sadegh H et al. The role of nanomaterials as effective adsorbents and their applications in wastewater treatment. Journal of Nanostructure in

Chemistry. 2017;**7**(1):1-14. DOI: 10.1007/s40097-017-0219-4

[15] Ali I. New generation adsorbents for water treatment. Chemical Reviews. 2012;**112**(10):5073-5091. DOI: 10.1021/cr300133d

[16] Raval NP, Shah PU, Shah NK. Adsorptive removal of nickel(II) ions from aqueous environment: A review. Journal of Environmental Management. 2016;**179**:1-20. DOI: 10.1016/j.jenvman.2016.04.045

[17] Valentin R. Agricultural waste-based nanomaterials for water purification. Aquananotechnology. 2021:577-595. DOI: 10.1016/B978-0-12-821141-0.00013-6

[18] Gu S, Kang X, Wang L, Lichtfouse E, Wang C. Clay mineral adsorbents for heavy metal removal from wastewater: A review. Environmental Chemistry Letters. 2019;**17**(2):629-654. DOI: 10.1007/s10311-018-0813-9

[19] Guan H, Zhao Y. Decontamination application of nanoclays. In: Clay Nanoparticles. Elsevier; 2020. pp. 203-224. DOI: 10.1016/B978-0-12-816783-0.00009-8

[20] Awad AM et al. Adsorption of organic pollutants by natural and modified clays: A comprehensive review. Separation and Purification Technology. 2019;**228**:115719. DOI: 10.1016/j.seppur.2019.115719

[21] Sarkar B, Rusmin R, Ugochukwu UC, Mukhopadhyay R, Manjaiah KM. Modified clay minerals for environmental applications. In: Modified Clay and Zeolite Nanocomposite Materials. Elsevier; 2019. pp. 113-127. DOI: 10.1016/B978-0-12-814617-0.00003-7

[22] Apreutesei RE, Catrinescu C, Teodosiu C. Surfactant-modified natural zeolites for environmental applications in water purification. Environmental Engineering and Management Journal. 2008;**7**(2):149-161. DOI: 10.30638/eemj.2008.025

[23] Sweetman M et al. Activated carbon, carbon nanotubes and graphene: Materials and composites for advanced water purification. C. 2017;**3**(4):18. DOI: 10.3390/c3020018

[24] Jlassi K, Krupa I, Chehimi MM. Overview. In: Clay-Polymer Nanocomposites. Elsevier; 2017. pp. 1-28. DOI: 10.1016/B978-0-323-46153-5.00001-X

[25] Han H, Rafiq MK, Zhou T, Xu R, Mašek O, Li X. A critical review of clay-based composites with enhanced adsorption performance for metal and organic pollutants. Journal of Hazardous Materials. 2019;**369**:780-796. DOI: 10.1016/j.jhazmat.2019.02.003

[26] Mao S, Gao M. Functional organoclays for removal of heavy metal ions from water: A review. Journal of Molecular Liquids. 2021;**334**:116143. DOI: 10.1016/j.molliq.2021.116143

[27] Bhattacharyya KG, Gupta SS. Kaolinite, montmorillonite, and their modified derivatives as adsorbents for removal of Cu(II) from aqueous solution. Separation and Purification Technology. 2006;**50**(3):388-397. DOI: 10.1016/j.seppur.2005.12.014

[28] Guégan R. Organoclay applications and limits in the environment. Comptes Rendus Chimie. 2019;**22**(2-3):132-141. DOI: 10.1016/j.crci.2018.09.004

[29] Bagherifam S et al. Highly selective removal of nitrate and perchlorate by organoclay. Applied Clay Science. 2014;**95**:126-132. DOI: 10.1016/j.clay.2014.03.021

[30] Xi Y, Mallavarapu M, Naidu R. Preparation, characterization of surfactants modified clay minerals and nitrate adsorption. Applied Clay Science. 2010;**48**(1-2):92-96. DOI: 10.1016/j.clay.2009.11.047

[31] Lee SM, Tiwari D. Organo and inorgano-organo-modified clays in the remediation of aqueous solutions: An overview. Applied Clay Science. 2012;**59-60**:84-102. DOI: 10.1016/j.clay.2012.02.006

[32] Williams LB, Haydel SE. Evaluation of the medicinal use of clay minerals as antibacterial agents. International Geology Review. 2010;**52**(7-8):745-770. DOI: 10.1080/00206811003679737

[33] Scano A, Cabras V, Pilloni M, Ennas G. Microemulsions: The renaissance of ferrite nanoparticle synthesis. Journal of Nanoscience and Nanotechnology. 2019;**19**(8):4824-4838. DOI: 10.1166/jnn.2019.16876

[34] de Paiva LB, Morales AR, Valenzuela Díaz FR. Organoclays: Properties, preparation and applications. Applied Clay Science. 2008;**42**(1-2):8-24. DOI: 10.1016/j.clay.2008.02.006

[35] Sato T. Effects of layer charge, charge location, and energy change on expansion properties of Dioctahedral Smectites. Clays and Clay Minerals. 1992;**40**(1):103-113. DOI: 10.1346/CCMN.1992.0400111

[36] Oyanedel-Craver VA, Fuller M, Smith JA. Simultaneous sorption of benzene and heavy metals onto two organoclays. Journal of Colloid and Interface Science. 2007;**309**(2):485-492. DOI: 10.1016/j.jcis.2006.10.001

[37] He H et al. Changes in the morphology of organoclays with HDTMA+ surfactant loading. Applied Clay Science. 2006;**31**(3-4):262-271. DOI: 10.1016/j.clay.2005.10.011

[38] Hamdi N, Srasra E. Acid-base properties of organosmectite in aqueous suspension. Applied Clay Science. 2014;**99**:1-6. DOI: 10.1016/j.clay.2014.07.009

[39] Msadok I, Hamdi N, Rodríguez MA, Ferrari B, Srasra E. Synthesis and characterization of Tunisian organoclay: Application as viscosifier in oil drilling fluid. Chemical Engineering Research and Design. 2020;**153**:427-434. DOI: 10.1016/j.cherd.2019.11.010

[40] Gammoudi S, Frini-Srasra N, Srasra E. Influence of exchangeable cation of smectite on HDTMA adsorption: Equilibrium, kinetic and thermodynamic studies. Applied Clay Science. 2012;**69**:99-107. DOI: 10.1016/j.clay.2011.11.011

[41] Xi Y. Synthesis, Characterisation and Application of Organoclays. Thesis Queensland University of Technology School of Physical and Chemical Inorganic Material Research Group. 2006. p. 85

[42] Shirzad-Siboni M, Khataee A, Hassani A, Karaca S. Preparation, characterization and application of a CTAB-modified nanoclay for the adsorption of an herbicide from aqueous solutions: Kinetic and equilibrium studies. Comptes Rendus Chimie. 2015;**18**(2):204-214. DOI: 10.1016/j.crci.2014.06.004

[43] Pandey P, De N. Surfactant-induced changes in physicochemical characters of bentonite clay. IRJPAC. 2018;**15**(4):1-11. DOI: 10.9734/IRJPAC/2017/39374

[44] De Oliveira T, Guégan R. Coupled organoclay/micelle action for the adsorption of diclofenac. Environmental

Science & Technology. 2016;**50**(18):10209-10215. DOI: 10.1021/acs.est.6b03393

[45] Gammoudi S, Frini-Srasra N, Srasra E. Preparation, characterization of organosmectites and fluoride ion removal. International Journal of Mineral Processing. 2013;**125**:10-17. DOI: 10.1016/j.minpro.2013.09.003

[46] Ma J, Cui B, Dai J, Li D. Mechanism of adsorption of anionic dye from aqueous solutions onto organobentonite. Journal of Hazardous Materials. 2011;**186**(2-3):1758-1765. DOI: 10.1016/j.jhazmat.2010.12.073

[47] Guégan R, Le Forestier L. Performance evaluation of organoclays for the amoxicillin retention in a dynamic context. Chemical Engineering Journal. 2021;**406**:126859. DOI: 10.1016/j.cej.2020.126859

[48] Ma J, Zhu L. Simultaneous sorption of phosphate and phenanthrene to inorgano–organo-bentonite from water. Journal of Hazardous Materials. 2006;**136**(3):982-988. DOI: 10.1016/j.jhazmat.2006.01.046

[49] Tabak A, Baltas N, Afsin B, Emirik M, Caglar B, Eren E. Adsorption of reactive red 120 from aqueous solutions by cetylpyridinium-bentonite. Journal of Chemical Technology and Biotechnology. 2010;**85**(9):1199-1207. DOI: 10.1002/jctb.2416

[50] Gamoudi S, Srasra E. Adsorption of organic dyes by HDPy+−modified clay: Effect of molecular structure on the adsorption. Journal of Molecular Structure. 2019;**1193**:522-531. DOI: 10.1016/j.molstruc.2019.05.055

[51] Behnsen J, Riebe B. Anion selectivity of organobentonites. Applied Geochemistry. 2008;**23**(9):2746-2752. DOI: 10.1016/j.apgeochem.2008.06.019

[52] Yahya K, Msadok I, Mlayah A, Srasra E, Hamdi N. Retention and selectivity of phosphate and fluoride from single and industrial aqueous solutions using purified and surfactant modified Tunisian clay. DWT. 2021;**216**:199-210. DOI: 10.5004/dwt.2021.26814

[53] Sharafimasooleh M, Bazgir S, Tamizifar M, Nemati A. Adsorption of hydrocarbons on modified nanoclays. IOP Conference Series: Materials Science and Engineering. 2011;**18**(18):182012. DOI: 10.1088/1757-899X/18/18/182012

[54] Saitoh T, Shibayama T. Removal and degradation of β-lactam antibiotics in water using didodecyldimethylammonium bromide-modified montmorillonite organoclay. Journal of Hazardous Materials. 2016;**317**:677-685. DOI: 10.1016/j.jhazmat.2016.06.003

[55] Yang J et al. Technetium-99 decontamination from radioactive wastewater by modified bentonite: Batch, column experiment and mechanism investigation. Chemical Engineering Journal. 2022;**428**:131333. DOI: 10.1016/j.cej.2021.131333

[56] Dultz S. Organophilic bentonites as adsorbents for radionuclides II. Chemical and mineralogical properties of HDPy-montmorillonite. Applied Clay Science. 2000;**16**(1-2):15-29. DOI: 10.1016/S0169-1317(99)00042-3

[57] Park Y, Ayoko GA, Kurdi R, Horváth E, Kristóf J, Frost RL. Adsorption of phenolic compounds by organoclays: Implications for the removal of organic pollutants from aqueous media. Journal of Colloid and Interface Science. 2013;**406**:196-208. DOI: 10.1016/j.jcis.2013.05.027

[58] Li D, Kaplan DI, Sams A, Powell BA, Knox AS. Removal capacity and chemical

speciation of groundwater iodide (I−) and iodate (IO3−) sequestered by organoclays and granular activated carbon. Journal of Environmental Radioactivity. 2018;**192**:505-512. DOI: 10.1016/j.jenvrad.2018.08.008

[59] Younker JM, Walsh ME. Impact of salinity and dispersed oil on adsorption of dissolved aromatic hydrocarbons by activated carbon and organoclay. Journal of Hazardous Materials. 2015;**299**:562-569. DOI: 10.1016/j.jhazmat.2015.07.063

[60] Jin X, Jiang M, Du J, Chen Z. Removal of Cr(VI) from aqueous solution by surfactant-modified kaolinite. Journal of Industrial and Engineering Chemistry. 2014;**20**(5):3025-3032. DOI: 10.1016/j.jiec.2013.11.038

[61] Yahya K, Msadok I, Mlayah A, Srasra E, Hamdi N. Retention and selectivity of phosphate and fluoride from single and industrial aqueous solutions using purified and surfactant modified Tunisian clay. Desalination and Water Treatment. 2021;**216**:13

[62] Hongping H, Ray FL, Jianxi Z. Infrared study of HDTMA+ intercalated montmorillonite. Spectrochimica Acta Part A: Molecular and Biomolecular Spectroscopy. 2004;**60**(12):2853-2859. DOI: 10.1016/j.saa.2003.09.028

[63] Ouellet-Plamondon CM, Stasiak J, Al-Tabbaa A. The effect of cationic, non-ionic and amphiphilic surfactants on the intercalation of bentonite. Colloids and Surfaces A: Physicochemical and Engineering Aspects. 2014;**444**:330-337. DOI: 10.1016/j.colsurfa.2013.12.032

[64] Zhang L, Zhang B, Wu T, Sun D, Li Y. Adsorption behavior and mechanism of chlorophenols onto organoclays in aqueous solution. Colloids and Surfaces A: Physicochemical and Engineering Aspects. 2015;**484**:118-129. DOI: 10.1016/j.colsurfa.2015.07.055

[65] Anirudhan TS, Ramachandran M. Adsorptive removal of basic dyes from aqueous solutions by surfactant modified bentonite clay (organoclay): Kinetic and competitive adsorption isotherm. Process Safety and Environmental Protection. 2015;**95**:215-225. DOI: 10.1016/j.psep.2015.03.003

[66] Krishna BS, Murty DSR, Jai Prakash BS. Surfactant-modified clay as adsorbent for chromate. Applied Clay Science. 2001;**20**(1-2):65-71. DOI: 10.1016/S0169-1317(01)00039-4

[67] Razmi B, Ghasemi-Fasaei R. Investigation of Taguchi optimization, equilibrium isotherms, and kinetic modeling for phosphorus adsorption onto natural zeolite of clinoptilolite type. Adsorption Science & Technology. 2018;**36**(7-8):1470-1483. DOI: 10.1177/0263617418779738

[68] Polubesova T, Zadaka D, Groisman L, Nir S. Water remediation by micelle–clay system: Case study for tetracycline and sulfonamide antibiotics. Water Research. 2006;**40**(12):2369-2374. DOI: 10.1016/j.watres.2006.04.008

[69] Birgersson M, Karnland O. Ion equilibrium between montmorillonite interlayer space and an external solution—Consequences for diffusional transport. Geochimica et Cosmochimica Acta. 2009;**73**(7):1908-1923. DOI: 10.1016/j.gca.2008.11.027

[70] Luo W, Ouyang J, Antwi P, Wu M, Huang Z, Qin W. Microwave/ultrasound-assisted modification of montmorillonite by conventional and gemini alkyl quaternary ammonium salts for adsorption of chromate and

phenol: Structure-function relationship. Science of the Total Environment. 2019;**655**:1104-1112. DOI: 10.1016/j.scitotenv.2018.11.329

[71] Hu B, Luo H. Adsorption of hexavalent chromium onto montmorillonite modified with hydroxyaluminum and cetyltrimethylammonium bromide. Applied Surface Science. 2010;**257**(3):769-775. DOI: 10.1016/j.apsusc.2010.07.062

[72] Seliem MK et al. Removal of nitrate by synthetic organosilicas and organoclay: Kinetic and isotherm studies. Separation and Purification Technology. 2013;**110**:181-187. DOI: 10.1016/j.seppur.2013.03.023

[73] Atia A. Adsorption of chromate and molybdate by cetylpyridinium bentonite. Applied Clay Science. 2008;**41**(1-2):73-84. DOI: 10.1016/j.clay.2007.09.011

[74] Gammoudi S, Frini-Srasra N, Srasra E. Nitrate sorption by organosmectites. Engineering Geology. 2012;**124**:119-129. DOI: 10.1016/j.enggeo.2011.10.009

Chapter 4

Solid-State Synthesis of Organoclays: Physicochemical Properties and Application for Bisphenol A Removal from Aqueous Solutions

Issaka Garikoé and Boubié Guel

Abstract

The solid-state intercalation of organoclays, by using surfactants bromide salts (n-$C_{12}H_{25}(CH_3)_3NBr$, n-$C_{14}H_{29}(CH_3)_3NBr$, n-$C_{16}H_{33}(CH_3)_3NBr$ and n-$C_{12}H_{25})_2(CH_3)_2NBr$) at different levels of the cation exchange capacity (CEC), was reported. According to XRPD data, the basal spacing of the smectite phase unit cell increased systematically up to >38 Å at 2.0-CEC of surfactants loading, indicating that they entered into the smectite interlayers and were arranged parallel to the layers at low concentrations and nearly vertical at high ones. Organoclays FTIR spectra showed a drop in the water band intensities at around 1630 and 3400 cm^{-1}, and new specific bands close to 2920 and 2850 cm^{-1} were assigned to the asymmetric and symmetric stretching of the surfactants CH_2 groups, respectively. Various factors, impacting bisphenol A (BPA) removal from water, were estimated in batch studies. It is shown that both physisorption and chemisorption describe the BPA sorption, and the pseudo-second-order kinetics and Langmuir isotherm fit better the BPA sorption results. The adsorption capacities of the organoclays are ranged from 88 to 127 mg/g at ambient temperature and in acidic solution. The effect of pH on the adsorption showed a remarkable drop in the quantity of adsorbed BPA when the pH is greater than 10 ($pH \geq 10$).

Keywords: organosmectites, surfactant, bisphenol A, adsorption, batch studies

1. Introduction

Endocrine disrupting chemicals (EDCs) are chemical pollutants that can affect the endocrine (hormonal) system and intrude on important developmental processes in humans and wildlife (World Health Organization, 1996). EDCs include natural estrogens (estrone: E1, 17β-estradiol: E2, estriol: E3), synthesized estrogen (ethinylestradiol: EE2), and artificial composites (bisphenol A: BPA, polychlorinated biphenyls: PCBs, polybrominated biphenyls: PBBs, and dioxins). EDCs are dichlorodiphenyltrichloroethane (DDT) from pesticides and vinclozolin from fungicides [1, 2]. Among these chemicals that are cited, BPA is produced in high volumes than the other ones and it can be found worldwide. Centers for Disease Control reported that BPA is present in the urine of 92.6% of over 2500 Americans [3, 4] and this indicates a general exposure to this chemical. Stockholm Convention has regulated the manufacturing and the utilization of DDT, PCBs, PBBs, and dioxins (Stockholm Convention on Persistent Organic Pollutants, 2009), and this agreement limited the exposition to these composites compared to BPA which is not regulated in all countries. A lot of studies reported that there was arising substantiation for adverse reproductive issues (fruitfulness, cancers, deformations) from exposure to EDCs, and there was also mounting substantiation of these chemicals on thyroid function, brain function, obesity and metabolism, and insulin and glucose homeostasis (Stockholm Convention on Persistent Organic Pollutants, 2009). Former studies reported the presence of BPA in wastewater, surface and ground waters, and indeed drinking water [5]. This urged major enterprises over endocrine-disrupting chemicals by scientists in recent years. The pollution of the environment by these chemicals is mainly via industrial activities [5]. Predominantly, it's used in the manufacturing of polycarbonate plastics and epoxy resins [6]. Also, dental securities, foods and potable barrels, polycarbonate tubes, cleansers, care products, etc., are listed as exposition sources (World Health Organization 1996). The United States Environmental Protection Agency has defined a daily dose of 50 mg of BPA/kg of body weight/day for BPA [6]. Although there is no database on BPA pollution in an aqueous environment in Burkina Faso, its presence could be supposed in wastewater, surface, and ground waters, due to numerous original factories and diligence whose products clearly lead to a general exposition of the population to BPA. The absence of legislation in many countries makes it urgent to develop simple and effective methods for BPA remediation in aqueous matrixes. Numerous procedures including adsorption, chemical advanced oxidation, membrane filtration, and electrochemical mineralization were utilized for BPA removal from water [2, 3, 7–13]. Among the listed procedures, adsorption seems to be appropriate due to its high effectiveness, ease of implementation, and low cost of the process due to the usefulness of various adsorbents. Numerous adsorbents including organo-montmorillonites [7, 8, 10–12, 14, 15], inorganic-organoclays [16], surfactant modified vermiculites [17], surfactant modified sepiolite [18], surfactant modified palygorskite [19], polymer based on kaolinite [20], chitosan [21], surfactant modified zeolite [2, 22, 23], graphene [9], etc., have been used for BPA removal from water.

Clay modification is widely done in aqueous solution via cation exchange or solid-state reactions. Organic molecules are inserted in dried clay interlayers by solid-state responses without washing with solvent and this makes the synthesis environmentally friendly and more favorable for industrialization [24]. Generally, the publication in this topic reported the organoclays synthesis using various procedures and chemicals, numerous type of clays, and diverse surfactants as reviewed elsewhere [24]. To obtain organoclays with hydrophobic properties, quaternary alkylammonium salts are the most frequently used, but n-alkyl-pyrrolidones, maleic anhydride, biomolecules, polymeric quaternary alkylammonium, alkyl-imidazolium, and phosphonium salts are also employed to the accomplishment of this outcome [24]. The field of application of organoclays is very large and this includes nanocomposites, adsorbents of organic and inorganic pollutants in soil, water, air, etc. [8, 24, 25]. Clays and organically modified clays are considerably employed in adsorption processes for the removal of organic pollutants due to their low cost and environmental friendly [3, 7, 8, 10, 21, 26].

Smectites are generally utilized due to their high cation exchange capacity (CEC), swelling aptitude, and high specific surface areas leading to high adsorption/absorption capacities [3, 7, 8, 10]. From the literature, it is known that organoclays employed for BPA removal from water are synthesized by liquid-state insertion of surfactants into the clay layers [16, 21, 26]. In the present investigations, the organoclays utilized to study the BPA sorption from water have been synthesized via a solid-state insertion methodology. The synthesis occurs via solid-state intercalation reaction by using two natural local clays (denoted AH and DI) and four cationic surfactants (dodecyl trimethylammonium, n-$C_{12}H_{25}(CH_3)_3N^+$ denoted C_{12}; tetradecyltrimethylammonium, n-$C_{14}H_{29}(CH_3)_3N^+$ denoted C_{14}; hexadecyltrimethylammonium, n-$C_{16}H_{33}(CH_3)_3N^+$ denoted C_{16} and didodecyldimethylammonium, (n-$C_{12}H_{25})_2(CH_3)_2N^+$ denoted $2C_{12}$).

The present work investigates: (i) the solid-state intercalation of three alkyltrimethyl-ammonium and one di-alkyldimethylammonium cation into two natural clays, and the subsequent changes in structure and physicochemical properties; (ii) the efficiency of organoclays prepared via solid-state methodology for BPA removal from water; (iii) various parameters such as kinetics (pseudo-first-order model and pseudo-second-order model), isotherm (Freundlich and Langmuir models) and thermodynamic parameters (standard enthalpy (ΔH), standard entropy (ΔS), and standard free energy (ΔG)) on BPA adsorption from water.

2. Characterization of synthesized organoclays

The organoclays synthesis and their characterization using various techniques have been published elsewhere [27].

2.1 X-ray diffraction of raw clays and organoclays

The **Figures 1a**, **b** and **2a–f** show the XRPD diffractograms of the AH and DI raw clays and the organoclays loaded at various loading levels (0.5 CEC, 1.0 CEC, 1.5 CEC and 2.0 CEC) of the C_{12}, C_{14}, C_{16}, and $2C_{12}$ surfactants. **Figure 1** shows that AH and DI clays contain quartz, montmorillonite, kaolinite, hematite, rutile, orthoclase, and anorthite only for DI clay.

From **Figures 1** and **2**, it can be seen that the d_{001} basal spacing raises when the surfactant loading level increases. This increase indicates that the surfactant molecules are inserted into the mineral clay layers. It is also seen that an increase in the length of the long alkyl chain (C_{12}, C_{14}, and C_{16}) and the number of long alkyl chains (C_{12} and $2C_{12}$) provokes an increase in the d_{001} value (**Figures 1** and **2**). Furthermore, the increase in the d_{001} values seems to take place in steps with preferred intervals, **Figures 1** and **2**.

As a matter of fact, the d_{001} basal spacing value of the organoclays loaded at 2.0 CEC is 25 Å for the C_{12} surfactant (2.0 CEC-C_{12}-AH and 2.0 CEC-C_{12}- DI) < 32 Å and 28 Å for the C_{14} surfactant (2.0 CEC-C_{14}-AH and 2.0 CEC-C_{14}-DI, respectively) < 38 Å for the C_{16} and $2C_{12}$ surfactants (2.0 CEC-C_{16}-AH and 2.0 CEC-C_{16}-DI, 2.0 CEC-$2C_{12}$-AH and 2.0 CEC-$2C_{12}$- DI). In a previous study, Park reported that at 0.5 CEC loading level, a marginal increase of the basal spacing (d_{001}) is observed when the long alkyl chain length increases: C_{12} (14.1 Å), C_{14} (14.3 Å), and C_{16} (14.4 Å) at a XRPD step size of 0.0167° for 2θ [28]. As a matter of principle, similar observation should be made in the present study for the d_{001} spacing for all prepared organoclays

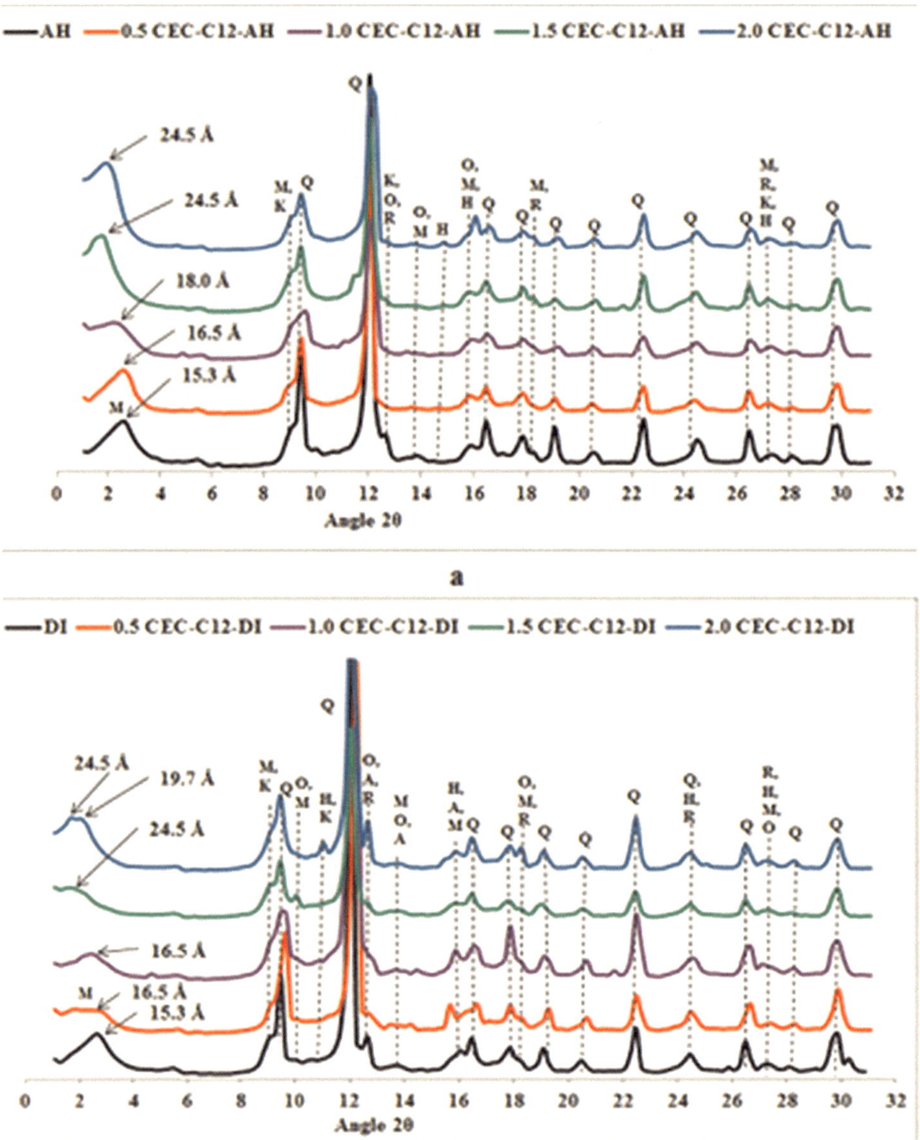

Figure 1.
X-ray powder diffraction patterns of AH, DI clays and C_{12}-AH, C_{12}-DI organoclays. Q = quartz, M = montmorillonite, K = kaolinite, H = hematite, R = rutile, O = orthoclase, and A = anorthite.

loaded at 0.5 CEC, **Figures 1** and **2**. This is related to the fact that at low loading level, the long alkyl chains are oriented parallel to the sheets, **Figures 3a** and **b**. The increase of the loading level, the length and the number of long alkyl chains in the quaternary alkylammonium cations increases the d_{001} basal spacing, **Figures 1a**, **b** and **2a–f**. The obtained results indicate that the orientation of the inserted quaternary alkylammonium cations varies from being parallel with the sheets to turn to more and more upright when the loading level increases (**Figure 3**).

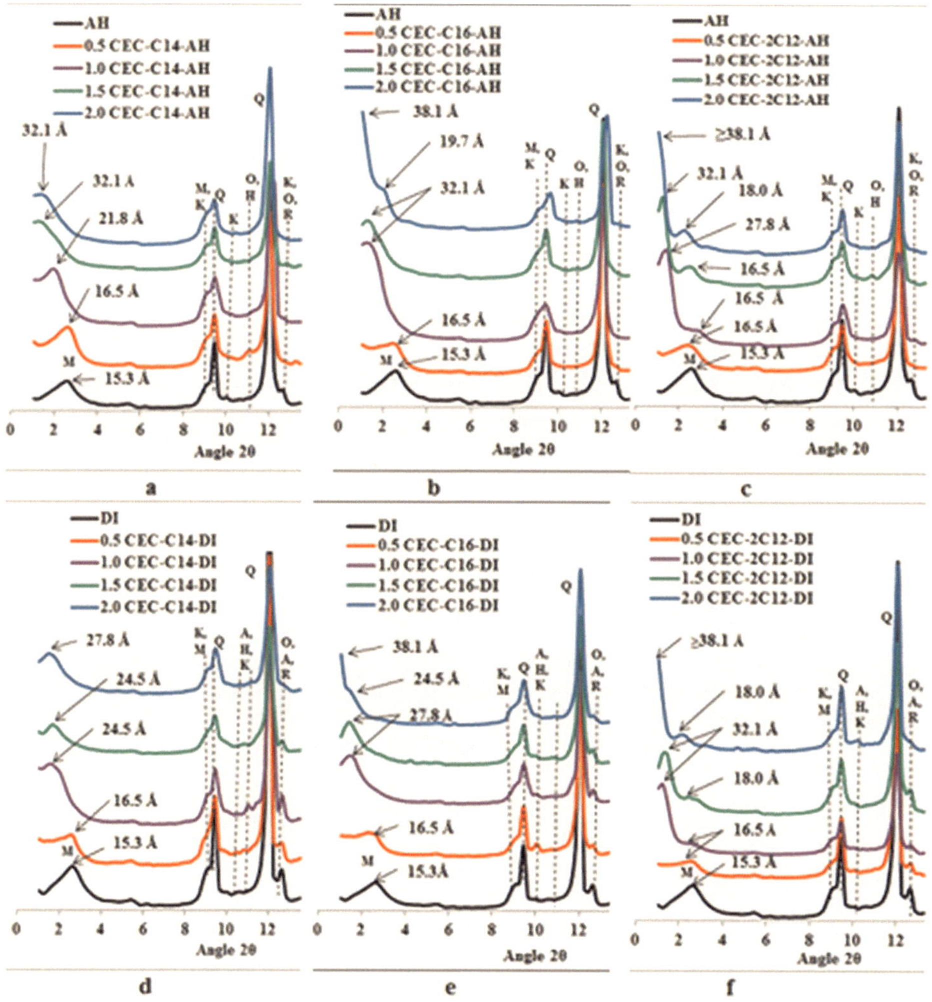

Figure 2.
X-ray powder diffraction patterns of AH, DI clays and C_{14}-AH, C_{14}-DI, C_{16}-AH, C_{16}-DI, $2C_{12}$-AH, and $2C_{12}$-DI organoclays. Q = quartz, M = montmorillonite, K = kaolinite, H = hematite, R = rutile, O = orthoclase, and A = anorthite.

The increase of the quaternary alkylammonium cations loading level from 0.5 CEC to 2.0 CEC levels causes the expansion of the montmorillonite interlayers to 6.8 Å, 8.3 Å, 10.0 Å, 12.1 Å, 14.8 Å, 18.1 Å, 22.4 Å and higher than 28.3 Å and these expansion values depend on the clays (AH or DI) and/or on the surfactants (C_{12}, C_{14}, C_{16}, and $2C_{12}$) used (**Figure 3**). Regarding the quaternary alkylammonium cations sizes, 8.3 Å and 10.0 Å as interlayer's expansions are attributed to a bilayer arrangement of the quaternary alkylammonium cations with an angle depending on the interaction between the quaternary alkylammonium cations and the clay layers, the chemical composition of the clay and the structure of the quaternary alkylammonium cations. This arrangement is expected for 18.0 Å and 19.7 Å as d_{001} basal spacing. The pseudo-trilayer arrangement is expected for 12.1 Å and 14.8 Å as basal spacing expansion and the long alkyl chain of the surfactants are orientated in

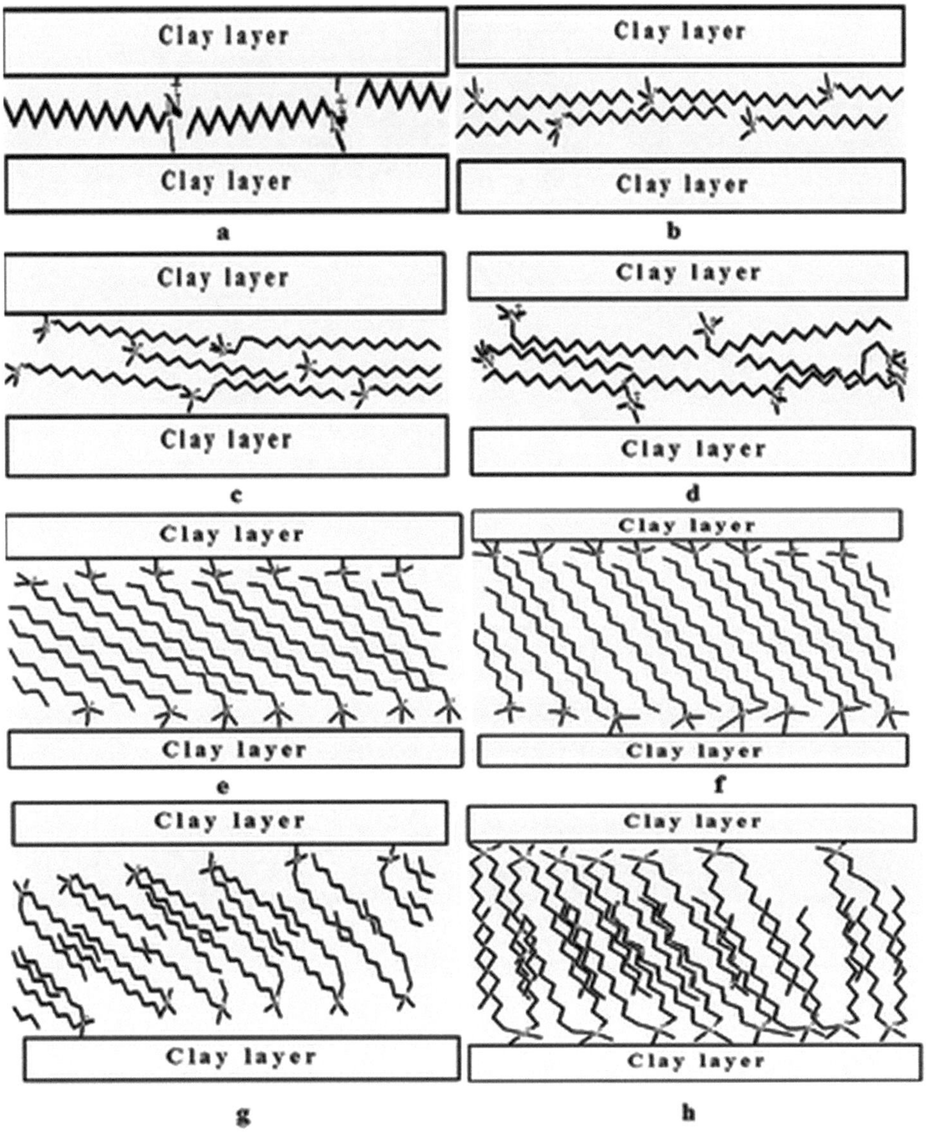

Figure 3.
Surfactant molecules orientations into the interlayer space: Arrangements (a) monolayer, (b) bilayers, and (c, d) pseudo-trilayers and (e, f and g, h) paraffin-type for one and two long alkyl chains respectively.

an oblique plan. For 18.1 Å, 22.4 Å and 28.4 Å as expansions, the long alkyl chain are arranged in a paraffin-type and their orientation approaches to the alternate anti-parallel packing in the clay interlayers as described in the crystal structures of the pure surfactants C_{12}, C_{14} and C_{16} [29]. This paraffin-type arrangement is expected when the d_{001} basal spacing values reach 27.8 Å, 32.1 Å or are higher than 38 Å. Therefore, the angle between the clay layers and the plane of the surfactant long alkyl chains approaches 90°. These proposed arrangements of the long alkyl chains in the organically modified clays interlayers are analogous to the arrangements reported by Lagaly et al. [30].

2.2 Fourier transform infrared (FTIR) spectroscopy of organoclays

Figures 4–7 represent the FTIR spectra of the raw clays, the organoclays loaded at 0.5 CEC, 1.0 CEC, 1.5 CEC and 2.0 CEC loading level of C_{12}, C_{14}, C_{16} and $2C_{12}$ surfactants, and pure surfactants.

The bending bands of the H-O appear at around 1600 cm^{-1} on the FTIR spectra of the raw and modified clays. The stretching vibrations of the structural hydroxyl (OH) is observed at 3622 cm^{-1} for the raw clays, and the band around 3420 cm^{-1} is attributed to the hydroxyl of water molecules contained in the clay layers. According

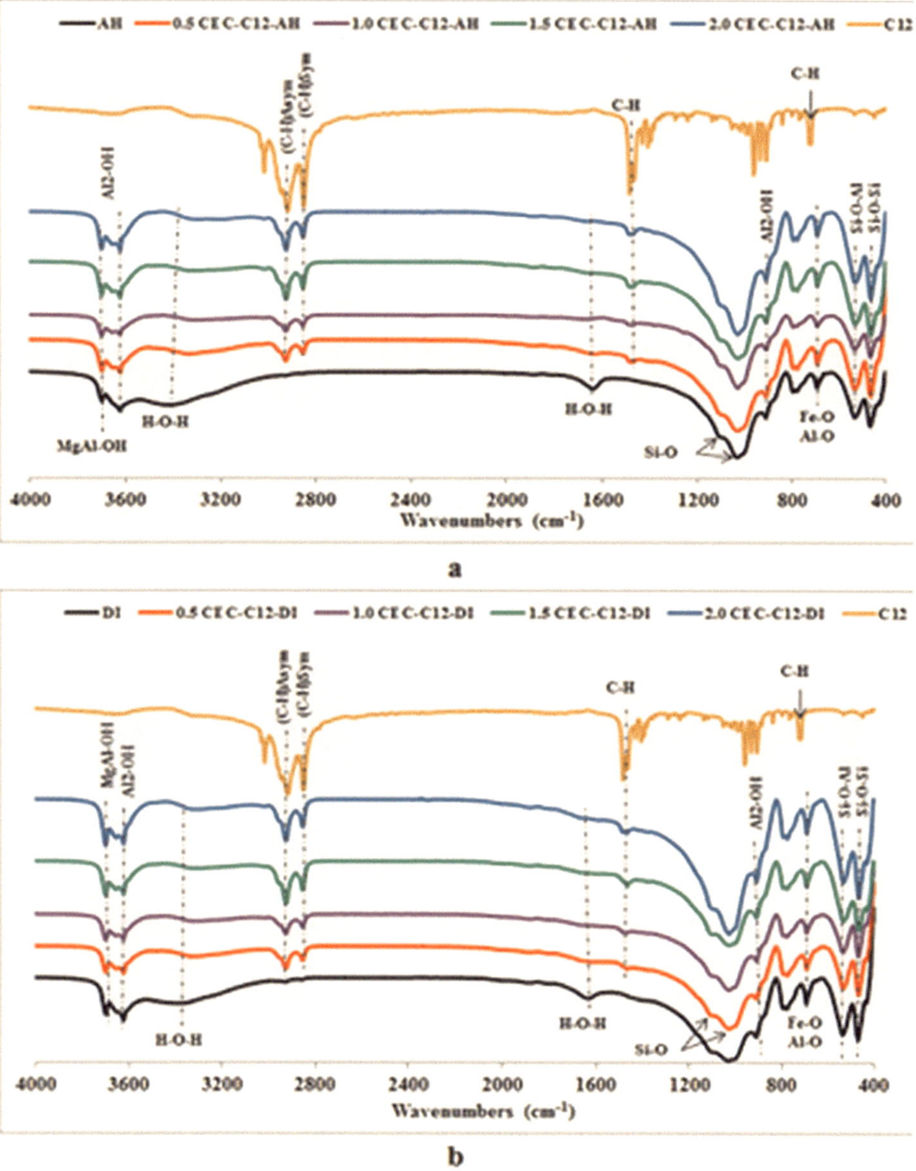

Figure 4.
FTIR spectra of AH and DI raw clays, pure C_{12} surfactant and C_{12}-AH and C_{12}-DI organoclays.

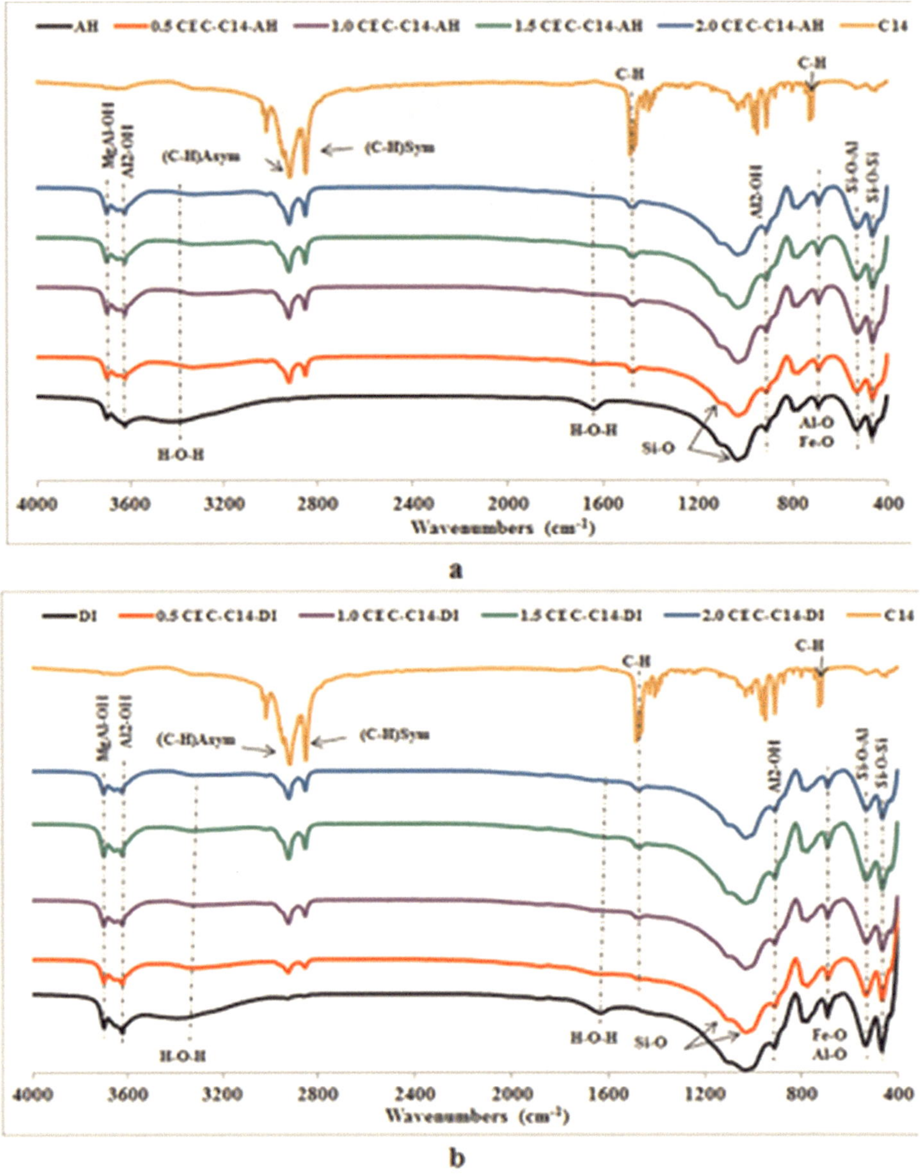

Figure 5.
FTIR spectra of AH and DI raw clays, pure C_{14} surfactant and C_{14}-AH and C_{14}-DI organoclays.

to Madejová et al., the bands seen at around 3694 and 3620 cm^{-1} are associated to the stretching vibrations of the structural hydroxyl in AlMgO-H and Al_2O-H of montmorillonite or kaolinite, respectively [31, 32]. In addition, the band observed at 3620 cm^{-1} on the spectra of raw clays and oganoclays are characteristic of smectites containing high amount of Al in the octahedra sheet [31]. The Si-O stretching band is observed at 1100 cm^{-1} and the Si-O in-plane bending appears at 470 cm^{-1} in all FTIR spectra [32]. The Al_2OH bending vibration and the Al-O-Si deformations bands are observed at 913 and 536 cm^{-1}, respectively [32]. The Fe-O or Al-O out of plane bending vibrations appear at around 694 cm^{-1} [33].

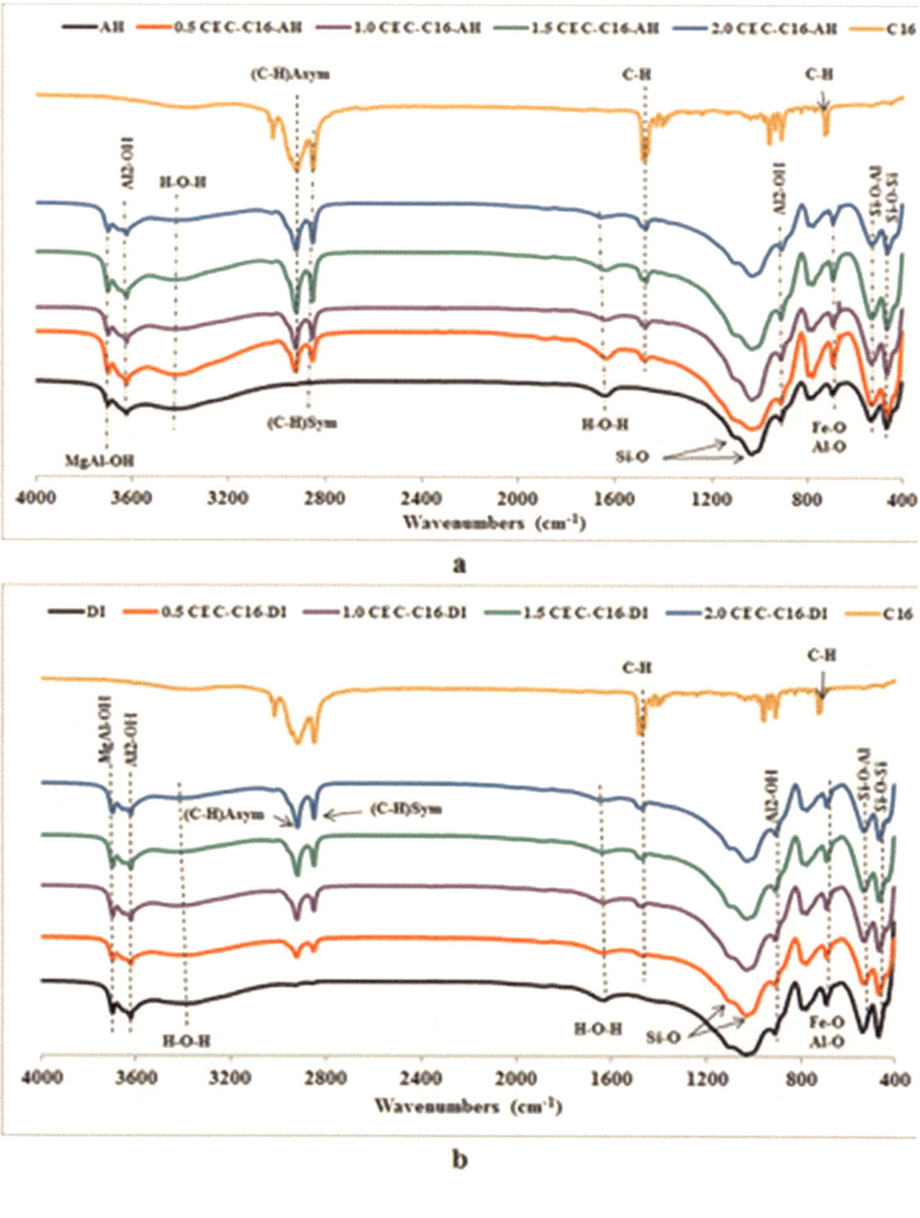

Figure 6.
FTIR spectra of AH and DI raw clays, pure C_{16} surfactant and C_{16}-AH and C_{16}-DI organoclays.

The asymmetric stretching bands of the CH_2 group shift slightly from 2927 to 2920 cm^{-1} depending on AH or DI raw clays, and surfactant loading, 0.5 to 2.0 CEC level due to the insertion of the C_{12}, C_{14}, $C_{16,}$ and $2C_{12}$ surfactants. Likewise, the wavenumbers of the CH_2 symmetric stretching vibrations shift from 2855 cm^{-1} to 2850 cm^{-1} with increasing surfactant loading levels (**Figures 4–7**). The wavenumbers of asymmetric and symmetric CH_2 stretching bands of surfactants in the modified clays increase with the surfactant loading level's increase and approach the wavenumbers of the pure surfactant. This further sustains that the long alkyl chains are packed identically in clay layers as in the pure quaternary alkylammonium

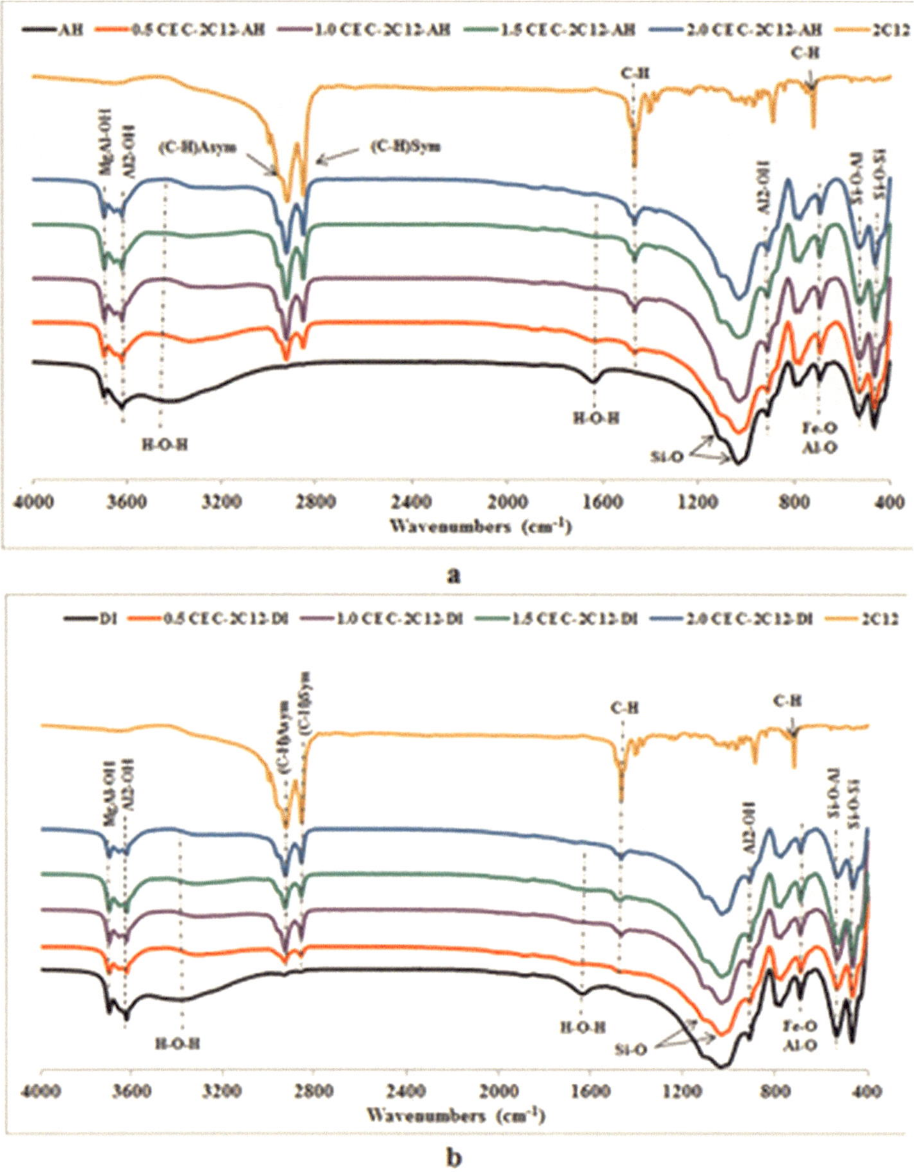

Figure 7.
FTIR spectra of AH and DI raw clays, pure $2C_{12}$ surfactant and $2C_{12}$-AH and $2C_{12}$-DI organoclays.

bromide salts when the loading level tends to saturation. In fact, Kamitori et al. [29], Silva et al. [34] and Campanelli and Scaramuzza [35] reported that the surfactants (C_{12}, $C_{14,}$ and C_{16}) are arranged in parallel plans inside a sheet and antiparallel as layers which alternate in the pure surfactant compound. The IR spectra of the organoclays support that the long alkyl chains orientation in the montmorillonite clay interlayers evolves progressively from a parallel orientation with clay sheets at lower loading level of surfactant to a standing position at higher loading level as reported in the literature [28, 36, 37].

	Surfactants and wavenumbers (cm^{-1}) of symmetric and asymmetric CH_2 stretching							
Loading level	**C_{12}**		**C_{14}**		**C_{16}**		**$2C_{12}$**	
	CH_2Sym	**CH_2Asym**	**CH_2Sym**	**CH_2Asym**	**CH_2Sym**	**CH_2Asym**	**CH_2Sym**	**CH_2Asym**
AC clay								
0.5 CEC	2855	2926	2853	2923	2852	2923	2854	2925
1.0 CEC	2854	2926	2852	2923	2851	2921	2854	2924
1.5 CEC	2854	2925	2852	2922	2851	2920	2853	2924
2.0 CEC	2854	2925	2851	2921	2850	2920	2853	2921
DI clay								
0.5 CEC	2855	2926	2854	2926	2852	2923	2855	2927
1.0 CEC	2855	2926	2853	2924	2851	2922	2855	2925
1.5 CEC	2854	2924	2852	2923	2851	2921	2854	2925
2.0 CEC	2854	2924	2852	2922	2851	2920	2854	2924
Pure surfactants	2850	2918	2849	2918	2849	2919	2853	2921

Table 1.
Change in CH_2 symmetric and asymmetric stretching vibration frequencies depending on the surfactant loading level and/or AH and DI clays.

The CH_2 asymmetric and symmetric stretching vibration bands frequencies in the pure surfactant molecules and the modified clays are summarized in **Table 1**. Contrary to the CH_2 asymmetric and symmetric stretching frequencies, the CH_2 scissoring and rocking vibrations frequencies observed at 1473 and 731 cm^{-1}, respectively [33], seem to be independent of the length of alkyl chain and level of loading.

Figures 4–7 show that the intensities of the band attributed to the hydroxyl of absorbed water seen at 3420 and 1635 cm^{-1} turn down progressively when the surfactant loading level increases and this behavior is certainly related to the change of the raw clay properties from hydrophilic to hydrophobic ones [33, 38]. The FTIR spectra support the model of insertion of the quaternary alkylammonium cations into mineral clay interlayers as described below, **Figure 3**.

3. Influence of various factors in batch studies of bisphenol A removal

In the present work, various factors such as the adsorbents dose (clays and organoclays), the agitation time, the initial concentration, the pH, the surfactants loading level, and the temperature have been studied in order to elucidate the BPA sorption process. The methodologies used for all adsorption experiences are described elsewhere [39].

3.1 Effect of the agitation time on the BPA adsorption and kinetics

3.1.1 Effect of the agitation time on the BPA adsorption

Figure 8a–d represent the amount of BPA removed as a function of the agitation time for 200 mg as raw clays and organoclays dose. From **Figure 8a–d**, the BPA

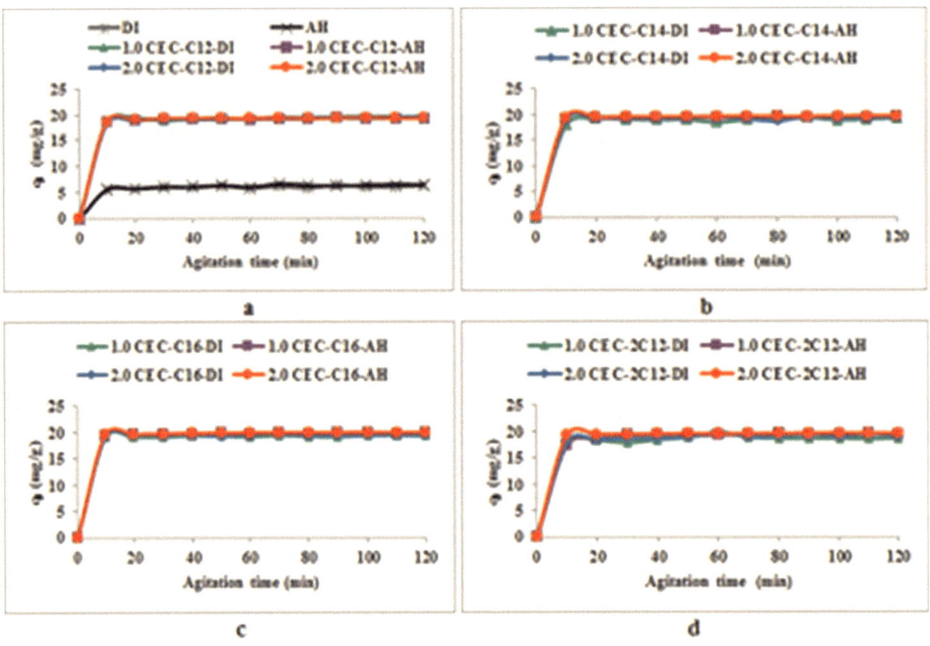

Figure 8.
Effect of agitation time of BPA adsorption on the raw clays and organoclays: AH, DI, C_{12}*-AH,* C_{12}*-DI (a),* C_{14}*-AH,* C_{14}*-DI (b),* C_{16}*-AH,* C_{16}*-DI (c) and* $2C_{12}$*-AH,* $2C_{12}$*-DI (d).*

sorption depends on the duration of the agitation. In fact, it is seen that the amount of BPA removed from the aqueous solution by the raw and modified clays increases during the first 10 minutes and attains the equilibrium concentration within 20 minutes (**Figure 8a–d**).

Moreover, after 20 minutes as duration of agitation, no significant difference is observed according to the types of surfactants, the surfactant loading levels and the clays types (**Figure 8a–d**). This finding is unlike to those reported by Zheng et al. [10], where the quantity of BPA removed per unit of mass increases with the surfactant chain longer (C_{12} to C_{16}). The agitation time required to reach the equilibrium is low and this suggests that the physisorption process is involved in the BPA removal [40]. It is also seen that the number of the surfactant long alkyl chain (C_{12} and $2C_{12}$) does not enhance the BPA adsorption and this results is unlike the results obtained by Park et al., where the quantity of BPA adsorbed increases with the increase of the number of long alkyl chain (C_{12} and $2C_{12}$) [8]. The ranking of the q_e values are from 18.7 to 19.8 mg of BPA/g of adsorbent for 200 mg as adsorbent dosage and 100 mg/L as BPA initial concentration for all the organoclays after 120 min as agitation time.

3.1.2 Kinetics

The constants of the pseudo-first and the pseudo-second kinetics have been evaluated by using the result of the effect of agitation time on BPA sorption and the obtained parameters are summarized into **Table 2** [39].

According to the high determination coefficients obtained ($R^2 > 0.98$) in the case of the pseudo-second order kinetic model when compared to those of the pseudo-

Clays or organoclays	Pseudo-first order			Pseudo-second order		
	K_1 (min^{-1})	q_{e1} (mg/g)	R^2	K_2 (min^{-1})	q_{e2} (g/mg.min)	R^2
AH	0.0145	0.958	0.6042	0.05	5.74	0.9831
DI	0.0065	0.683	0.2835	0.07	5.77	0.9871
1.0 CEC-C_{12}-AH	0.0053	0.701	0.5089	0.14	19.34	0.9999
2.0 CEC-C_{12}-AH	0.0062	0.427	0.3684	0.30	19.34	0.9999
1.0 CEC-C_{12}-DI	0.0064	1.055	0.5983	0.05	19.72	0.9998
2.0 CEC-C_{12}-DI	0.0110	0.412	0.6269	0.06	19.57	0.9999
1.0 CEC-C_{14}-AH	0.0053	0.701	0.6828	0.17	19.49	1.0000
2.0 CEC-C_{14}-AH	0.0062	0.427	0.7616	0.20	19.72	1.0000
1.0 CEC-C_{14}-DI	0.0064	1.055	0.2460	0.08	19.23	0.9993
2.0 CEC-C_{14}-DI	0.0110	0.412	0.2639	0.42	19.38	0.9991
1.0 CEC-C_{16}-AH	0.0225	0.554	0.2877	0.26	19.76	0.9998
2.0 CEC-C_{16}-AH	0.0330	0.571	0.7684	0.15	19.96	1.0000
1.0 CEC-C_{16}-DI	0.0031	0.854	0.2951	0.17	19.34	0.9998
2.0 CEC-C_{16}_DI	0.0308	1.263	0.4490	0.13	19.65	0.9998
1.0 CEC-2C_{12}-AH	0.0122	1.405	0.4870	0.05	19.69	0.9999
2.0 CEC-2C_{12}-AH	0.0062	0.598	0.8406	0.16	19.69	1.0000
1.0 CEC-2C_{12} -DI	0.0043	1.377	0.3742	0.13	18.76	0.9998
2.0 CEC-2C_{12} -DI	0.0112	0.848	0.4506	0.06	19.34	0.9989

Table 2.
Kinetic constants of BPA removal on raw and modified clays.

first-order kinetic model ones ($R^2 < 0.85$) (**Table 2**), it can be concluded that the removal of BPA is better described by the pseudo-second-order kinetic model than the pseudo-first order kinetic model (**Table 2**). According to the literature, the pseudo-second-order kinetic model suggests that the chemisorption is the limiting step during the adsorption process [8, 16]. This result is like those reported previously on BPA removal using different adsorbents, such as graphene [9], C_{16}-palygorskite [19], surfactant-modified zeolite [26], organo-montmorillonites [3, 7, 8, 10, 16]. Generally, BPA removal from water using organoclays is better described by the pseudo-secon-order kinetic than the pseudo-first order kinetic [7, 8, 10, 11, 16, 21].

3.2 Effect of pH on BPA adsorption

Figure 9 represents the amount of BPA removed when the pH increases from 2 to 12. It is seen that from acidic pH up to pH 8, the amount of BPA removed on all organoclays stays relatively constant (**Figure 9a-d**). For pH values ranging from 2 to 8, the q_e values ranged from 19.4 to 19.8 mg/g for all modified clays. This is followed by a decrease in the efficiency of organoclays in BPA sorption from aqueous solution having pH ≥ 8. In fact, the qe values decrease slightly from 19.8–19.4 mg/g to 19.6 – 16.6 mg/g when the pH value increases from 8 to 10. The q_e values turn down remarkably when the pH becomes higher than 10 for all the organoclays with higher

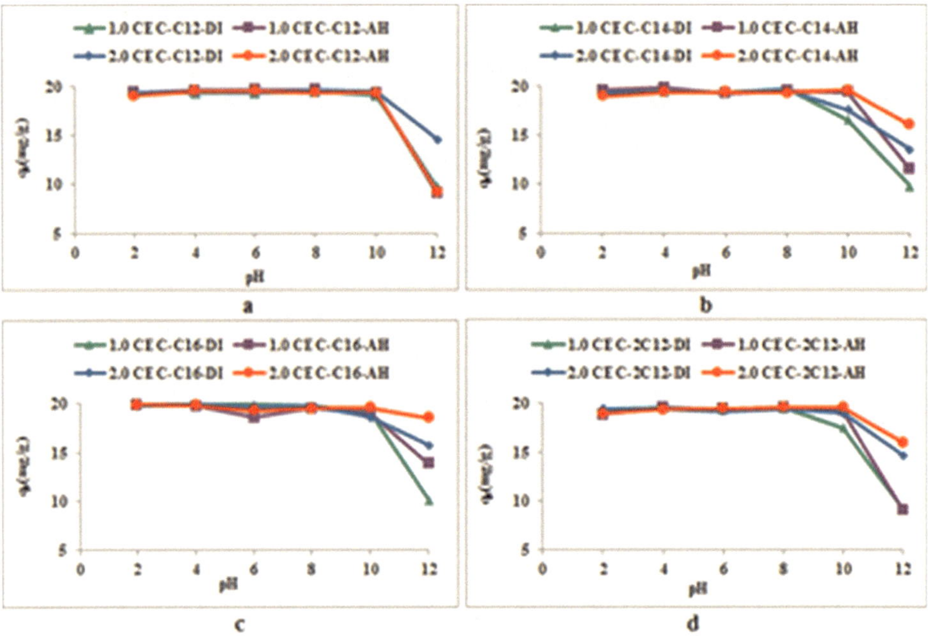

Figure 9.
Effect of pH on the adsorption of BPA by organoclays: C_{12}-AH and C_{12}-DI (a), C_{14}-AH and C_{14}-DI (b), C_{16}-AH and C_{16}-DI (c), $2C_{12}$-AH and $2C_{12}$-DI (d).

decrease for those loaded at 1.0 CEC. In fact, it is seen that the qe values for all modified clays are ranged between 18 to 9.0 mg /g when the pH varies from 10 to 12. For pH = 12, the organoclays loaded at 2.0 CEC loading level give higher q_e values (9.2 to 18.5 mg/g) compared to those loaded at 1.0 CEC (9.0 to 13.9 mg/g). In addition, higher q_e values are obtained for the organoclays based on AH clay loaded at 2.0 CEC than the ones based on DI clay loaded at 2.0 CEC and only for C_{14}, C_{16} and $2C_{12}$ surfactants at a pH = 12. An opposite effect is observed for C_{12} surfactant because the C_{12}-DI organoclays yielded high qe values than the C_{12}-AH ones at 2.0 CEC loading level. This result which is obtained in the case of high pH value (pH = 12) seems to be linked to the high ability of AH clay to insert more quaternary alkylammonium cations than the DI clay one, due to the fact that AH clay has a high CEC value than the DI clay. The BPA molecules ionization degree depends on the pH of the aqueous solution and this seems to be another reason for the decrease in the q_e values when the pH reached 8 [2, 8, 9]. In fact, BPA molecules remain predominantly in their neutral form when pH $\leq$ 8, in their mono-ionized form when $8 \leq pH \leq 10$ ($pKa_1 = 9.6$) and in the di-ionized form when pH $\geq$ 10 ($pKa_2 = 11.3$). Thus, the adsorption capacity's decrease of the modified clays seen in the case of alkaline pH range can be explained by the electrostatic repulsions existing between the negative charges of the clay's surface and those bisphenolate anions [9]. In the present work, the efficiency of organoclays to sorb BPA in alkaline solution (pH $\geq$8) is less influenced for organoclays loaded at 2.0 CEC than those ones loaded at 1.0 CEC. This result sustains that the modified clays loaded at 2.0 CEC maintain high affinity toward the mono-ionized and the di-ionized forms of BPA at high pH values. Rathnayake et al. reported a possible retention of surfactant molecules onto the clays surface, which provides sites having hydrophobic character for BPA removal [16]. **Figure 9** shows that in the case of a solution with pH ≤ 8, 1.0 CEC as loading level is sufficient to get organoclays having high removal capacities for BPA, whereas in the case of a solution with pH value ≥8, 2.0 CEC as loading level is required to get organoclays having high removal capacities toward BPA. Particularly, when the pH reaches 12, we recommend the employment of either 2.0 CEC-C_{14}-AH, 2.0 CEC-C_{16}-AH, 2.0 CEC-2C_{12}-AH organoclays, or 2.0 CEC-C_{12}-DI organoclay.

3.3 Effect of BPA initial concentration on BPA adsorption and isotherms

3.3.1 Effect of BPA initial concentration on BPA adsorption

Figure 10a-d represent the amount of BPA removed as a function of the initial concentration for 12 hours as agitation duration.

It is seen that the quantity of BPA removed from the aqueous solution increases rapidly up to 200 mg/L as initial concentration. **Figure 10a-d** show that the equilibrium adsorption capacity of BPA on modified clays increases when the initial concentration increases and this until to the saturation of available reactive sites [11]. As shown in **Figure 10a-d**, all curves on the effect of initial concentration on BPA adsorption onto the organoclays are almost relatively linear up to 200 mg/L and become parabolic after that value. As a result, the removal of BPA onto the organoclays is affected by its initial concentration. This can be explained by the fact that before 200 mg/L is BPA initial concentration, there are more available active sites onto the organoclays which causes a better elimination rate of BPA. However, the number of active sites decrease gradually after 200 mg/L and this induces less efficient adsorption. As reported in previous studies, the present work shows that the

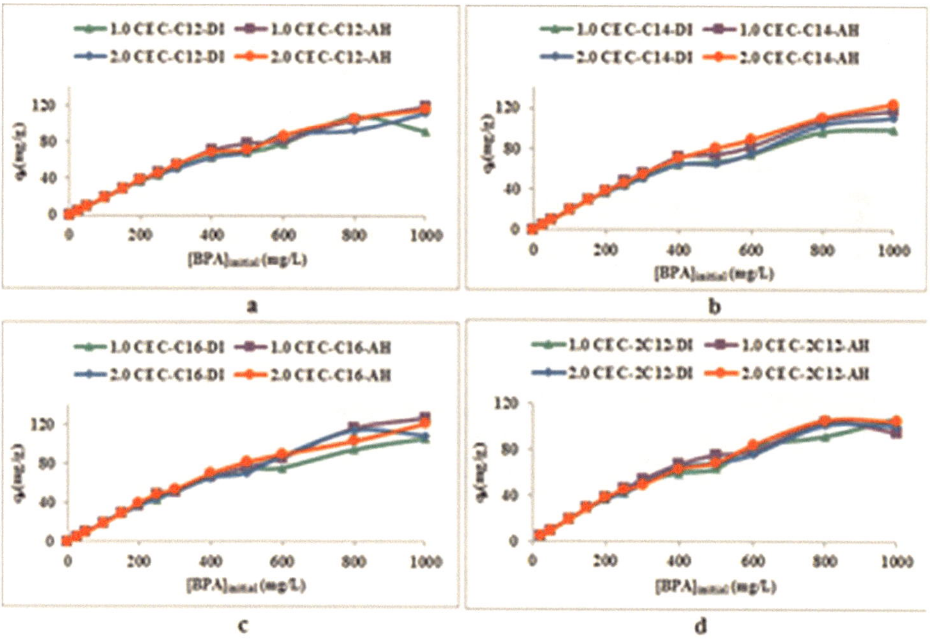

Figure 10.
Initial concentration effect on BPA adsorption by organoclays: C_{12}-AH and C_{12}-DI (a), C_{14}-AH and C_{14}-DI (b), C_{16}-AH and C_{16}-DI (c), 2C_{12}-AH and 2C_{12}-DI (d).

increase of BPA initial concentration causes an increase in the q_e value. For example, for 200 mg as adsorbent dose, when we increase the initial concentration from 25 to 1000 mg/L, the q_e values increase from 4.8–4.9 mg/g to 88–127 mg of BPA/g of adsorbent for all the organoclays.

3.3.2 Isotherms of BPA adsorption

The effect of the initial concentration has been used to evaluate the Langmuir and Freundlich isotherms parameters [39] and the obtained values are included in **Table 3**. **Table 3** shows that the determination coefficient (R^2) values are better in the case of the Langmuir isotherm ($0.89 \leq R^2 \leq 0.99$) than the Freundlich isotherm ($0.89 \leq R^2 \leq 0.95$) for all the organoclays and this corresponds to a monolayer adsorption of BPA. The above results resemble to those obtained on BPA adsorption using organically modified montmorillonite [7, 10] and amphoteric modified vermiculite [17].

3.4 Effect of surfactant loading level on BPA adsorption

The **Figure 11a-d** represent the qe values of adsorbed BPA as a function of the surfactant loading level (aqueous solution pH around 5–6). **Figure 11a–d** show high elimination rate of BPA onto AH and DI modified clays compared to the untreated ones. This result states that the intercalation of the quaternary alkylammonium into the clays' interlayers induces a strong removal level of BPA from aqueous solution.

Organoclays	Freundlich			Langmuir		
	K_F (L/g)	n	R^2	K_L (L/mg)	q_m (mg/g)	R^2
1.0 CEC-C_{12}-AH	13.2355	2.5588	0.9136	0.044	114.9	0.9700
2.0 CEC-C_{12}-AH	12.4460	2.5208	0.9172	0.047	108.7	0.9745
1.0 CEC-C_{12}-DI	10.9968	2.6199	0.9106	0.044	92.6	0.9654
2.0 CEC-C_{12}-DI	10.2492	2.4337	0.8963	0.019	113.6	0.8890
1.0 CEC-C_{14}-AH	12.0903	2.2188	0.9270	0.046	125.0	0.9756
2.0 CEC-C_{14}-AH	10.6440	2.2311	0.9464	0.050	125.0	0.9766
1.0 CEC-C_{14}-DI	11.0012	2.6100	0.9060	0.038	97.1	0.9635
2.0 CEC-C_{14}-DI	10.8786	2.2750	0.9100	0.036	96.1	0.9460
1.0 CEC-C_{16}-AH	12.1521	2.4384	0.9407	0.310	126.6	0.9379
2.0 CEC-C_{16}-AH	11.4857	2.3180	0.8853	0.044	119.1	0.9800
1.0 CEC-C_{16}-DI	10.5018	2.6157	0.8885	0.046	112.4	0.9579
2.0 CEC-C_{16}_DI	10.7585	2.4050	0.9280	0.034	111.1	0.9747
1.0 CEC-2C_{12}-AH	10.6867	2.4290	0.8920	0.051	99.0	0.9900
2.0 CEC-2C_{12}-AH	10.1757	2.3430	0.8881	0.035	108.7	0.9659
1.0 CEC-2C_{12} -DI	11.2105	2.6302	0.9215	0.060	87.7	0.9704
2.0 CEC-2C_{12}-DI	10.6515	2.5589	0.9482	0.031	103.1	0.9680

Table 3.
Isotherm parameters of BPA adsorption on the organoclays.

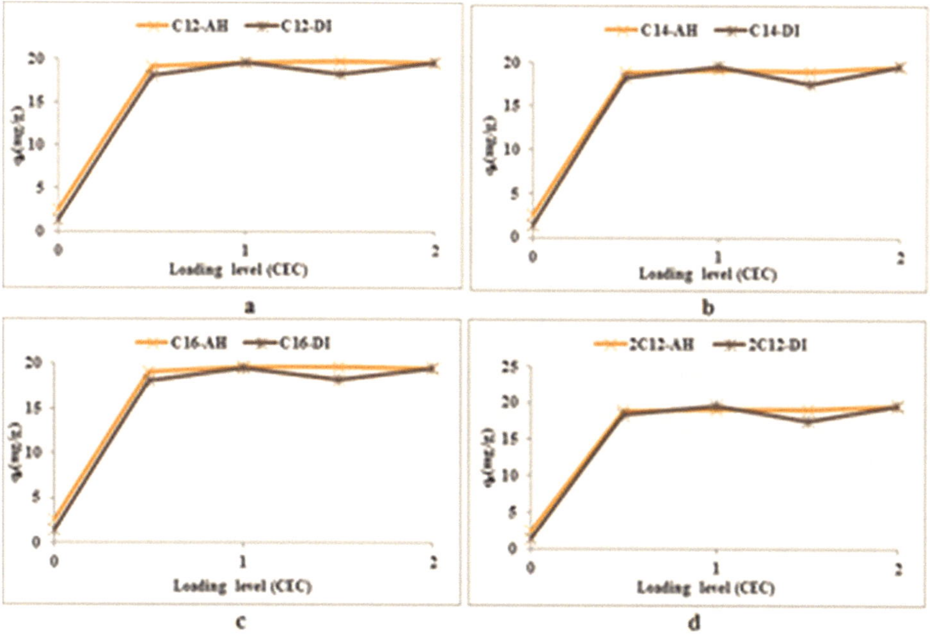

Figure 11.
The C_{12}, C_{14}, C_{16}, and $2C_{12}$ surfactants loading level effect on the BPA adsorption.

As a result of the surfactants intercalation, the amount of BPA sorbed per mass of adsorbent (q_e) increases remarkably from 1.4 and 2.5 mg/g for DI and AH raw clays, respectively, to q_e values higher than 19.5 mg/g in the case of organoclays loaded at 2.0 CEC loading level. We notice that the efficiency of the modified clays to remove BPA from aqueous solution is not significantly affected when the surfactant loading level increases from 1.0 to 2.0 CEC for pH $\leq$ 8. The below result is unlike those reported by Park et al. who found that the amount of BPA adsorbed increases when the surfactant loading level increases [8]. According to **Figure 11**, a better elimination rate is obtained in acidic solution when the AH and DI raw clays are loaded at least at 1.0 CEC loading level of the C_{12}, C_{14}, C_{16}, and $2C_{12}$ surfactants, which result is identical to the those obtained by studying the effect of the pH on the BPA sorption.

3.5 Effect of temperature on BPA adsorption and thermodynamics

3.5.1 Effect of temperature on BPA adsorption

The temperature effect on BPA adsorption onto the organoclays was evaluated at different temperatures (303, 313 and 323 K) and the obtained results are presented in **Figure 12a–d**.

Figure 12 shows that the increase of the temperature from 30 to 50°C causes a slight drop of the amount of BPA removed on the organoclays. This result shows that BPA removal is more propitious for lower temperatures. In fact, the ranking of the q_e values obtained are from 19.6 and 18.5 mg of BPA/g of organoclays and this occurs when the temperature increases from 30 to 50°C. This is due to the exothermic nature of the adsorption reaction [41, 42]. This means that the temperature affects weakly

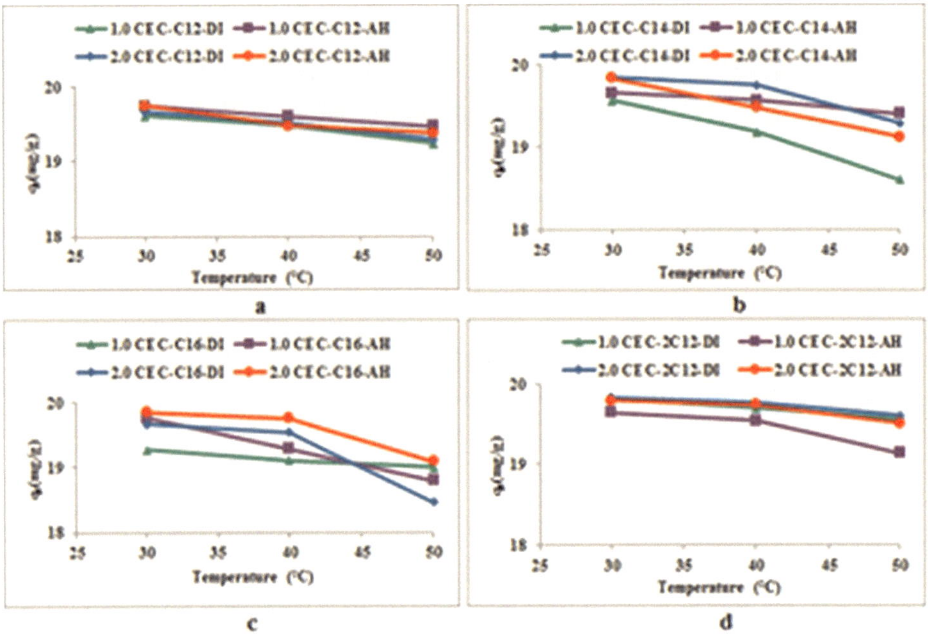

Figure 12.
Effect of the temperature on BPA adsorption by the organoclays: C_{12}-AH and C_{12}-DI (a), C_{14}-AH and C_{14}-DI (b), C_{16}-AH and C_{16}-DI (c), $2C_{12}$-AH and $2C_{12}$-DI (d).

the uptake of BPA onto the organoclays. The organoclays maintain high removal level toward BPA molecules at a temperature up to 50°C ($q_e \geq 18.5$ mg/g).

3.5.2 Thermodynamics

Figure 12a-d represent the quantities of BPA removed on the organoclays at different temperatures (303, 313, and 323 K). These figures show that the quantity of BPA removed on the organoclays drop weakly with the increase of the temperature from 30 to 50°C and this results means that the adsorption process is favored at lower temperature.

The results obtained by studying the effect of the temperature on BPA removal have been used to calculate the thermodynamic parameters such as Gibbs free energy (ΔG), enthalpies change (ΔH), and entropy change (ΔS) and the calculated values are included in **Table 4**. All enthalpies change values (**Table 4**) are less than zero ($\Delta H \leq 0$) and this suggests that the uptake of BPA from water using organoclays is exothermic, which is also sustained by the drop of the amount of BPA removed with the increase of the temperature. From the literature, it can be noted that when the values of ΔH are ranged from 2.1 to 20.9, a physisorption is involved in the adsorption process, but chemisorption intervenes for a value ranging from 80 to 200 kJ/mol [43]. In this work, the ΔH values evaluated from the experimental data of BPA sorption onto the organoclays are ranged from −76.5 to −13 kJ/mol. According to the ΔH values, different forces and interactions are supposed to be involved in the adsorption process [19]: between 4 and 10 kJ/mol for Van der Waals interactions, about 5 kJ/mol for hydrophobic interactions, between 2 and 40 kJ/mol for hydrogen bond

Organoclays	Thermodynamics constants					
	Temperature (K)	ΔG (kJ/mol)	R^2	ΔH (kJ/mol)	ΔS (J/mol.K)	R^2
1.0 CEC-C_{12}-AH	303	−8.651	0.9899	−31.382	−80.782	0.9830
	313	−8.936				
	323	−9.222				
2.0 CEC-C_{12}-AH	303	−8.525	0.9938	−38.429	−103.958	0.9978
	313	−8.806				
	323	−9.087				
1.0 CEC-C_{12}-DI	303	−7.557	0.9927	−27.847	−72.486	0.9838
	313	−7.806				
	323	−8.055				
2.0 CEC-C_{12}-DI	303	−7.861	0.9893	−31.537	−83.439	0.9976
	313	−8.120				
	323	−8.380				
1.0 CEC-C_{14}-AH	303	−8.035	0.9941	−23.714	−57.736	0.9927
	313	−8.300				
	323	−8.566				
2.0 CEC-C_{14}-AH	303	−8.445	0.9623	−62.330	−181.278	0.9777
	313	−8.724				
	323	−9.003				
1.0 CEC-C_{14}-DI	303	−6.734	0.9756	−49.269	−144.431	0.9993
	313	−6.956				
	323	−7.179				
2.0 CEC-C_{14}-DI	303	−9.018	0.9748	−64.258	−185.011	0.9950
	313	−9.316				
	323	−9.613				
1.0 CEC-C_{16}-AH	303	−7.897	0.9583	−69.123	−205.322	0.9666
	313	−8.158				
	323	−8.418				
2.0 CEC-C_{16}-AH	303	−8.854	0.9737	−76.408	−223.372	0.9989
	313	−9.147				
	323	−9.439				
1.0 CEC-C_{16}-DI	303	−6.454	0.9984	−13.375	−30.347	0.9811
	313	−6.666				
	323	−6.881				
2.0 CEC-C_{16}-DI	303	−7.216	0.9994	−48.506	−139.467	0.9996
	313	−7.454				
	323	−7.692				

Organoclays	Thermodynamics constants					
	Temperature (K)	ΔG (kJ/mol)	R^2	ΔH (kJ/mol)	ΔS (J/mol.K)	R^2
1.0 CEC-$2C_{12}$-AH	303	−7.514	0.9924	−36.103	−99.726	0.9989
	313	−7.762				
	323	−8.010				
2.0 CEC-$2C_{12}$-AH	303	−9.237	0.9940	−34.668	−90.032	0.9974
	313	−9.541				
	323	−9.846				
1.0 CEC-$2C_{12}$-DI	303	−9.177	0.9867	−34.165	−87.729	0.9911
	313	−9.480				
	323	−9.783				
2.0 CEC-$2C_{12}$-DI	303	−9.398	0.9883	−34.956	−87.305	0.9888
	313	−9.708				
	323	−10.018				

Table 4.
Thermodynamic constants of BPA adsorption on the modified clays.

interactions, around 40 kJ/mol for coordination exchange interactions, between 2 and 29 kJ/mol for dipole bond, and up to 60 kJ/mol for chemical bond. All ΔH values obtained in the present work are ranged from −76.5 to −13 kJ/mol and this suggests that the following interactions can intervene during the adsorption process: dipole-bonding, coordination exchange, hydrogen bonding and chemical bond. From the lower adsorption efficiency of the raw clays toward BPA molecules, we suggest that the hydrogen bonding contributes weakly to the BPA sorption in the present work [11]. From the obtained results, it can be postulated that both physisorption (Van der Waals forces, hydrogen bonding) and chemisorption (hydrophobic interactions, BPA phenolic rings and alkylammonium group interactions) intervene during BPA sorption onto the organically modified clays. The results obtained in the present work resemble to those reported previously on the adsorption of Rhodamine-B onto activated carbon [44]; adsorption of Acid Red G dye onto C_{18}-montmorillonite [45], adsorption of carbofuran onto biochars [46]. According to previous reported results, adsorption can be due to the intervention of both physisorption and chemisorption and this coexistence is described as the following: (i) a first rapid chemisorption followed by a slow physisorption [47]; (ii) contributions of both processes [48]; (iii) simultaneous occurring of both physisorption and chemisorption with predominant chemisorption [46] or without any predominant process [49].

Table 4 indicates that $\Delta G \leq 0$ and this suggests that the uptake of BPA molecules by the organoclays is spontaneous. The drop of the ΔG values with the increase of the temperature shows that the capacity of the organoclays to sorb BPA molecules is inversely dependent on the temperature [9, 19, 50], which also indicates a spontaneous process at low temperatures. In addition, ΔG decreases with the increase of the temperature (**Table 4**) and this shows that BPA sorption is thermodynamically favored at room temperature than at higher temperatures [51]. Gibbs free energy ranging from −20 to 0 kJ/mol corresponds to physisorption mechanism, while values ranging from -80 to -400 kJ/mol suggest chemisorption mechanism [52]. In the present work, the Gibbs free-energy values are ranged from -11 to -6 kJ/mol (**Table 4**) and this result means that the physisorption intervenes in BPA removal from water. According to previous literature results, when ΔS value is higher than zero ($\Delta S > 0$) the interactions at the solid–liquid interface are aleatory, while value less than zero ($\Delta S < 0$) suggests inverse interactions [50, 53]. In the present work, the ΔS values are all negative ($-30 \leq \Delta S \leq -205$ J/mol) and this suggests a decrease of the disorder at the solid–liquid interface, while ΔG and ΔH negative values indicate spontaneous and exothermic processes of BPA sorption onto the organoclays [8].

Our study highlights the fact that the adsorption is spontaneous ($\Delta G < 0$), exothermic ($\Delta H < 0$) with negative entropy values ($\Delta S < 0$) and these results are in agreement with those previously reported on the adsorption of BPA using organoclays [7, 8, 10], C_{16}-modified palygorskite [19], and surfactant-modified zeolite [22, 54].

4. Conclusion

The solid-state reaction methodology was successfully used to intercalate three cationic alkyltrimethylammoniums ($n\text{-}C_{12}H_{25}(CH_3)3\ N^+$, $n\text{-}C_{14}H_{29}(CH_3)_3N^+$ and $n\text{-}C_{16}H_{33}(CH_3)_3N^+$) and one di-alkyldimethylammonium ($(n\text{-}C_{12}H_{25})_2(CH_3)_2N^+$) into two natural clays containing Ca-montmorillonite as main clay type. XRPD shows that the d_{001} basal spacing increases after the intercalation of the cationic surfactants and this proved the swelling of the clays layers. It is observed that at 2.0 CEC as surfactant loading level, the basal spacing (d_{001}) of the prepared organoclays are 25 Å and 32 Å for those treated with C_{12} and C_{14} surfactants respectively and up to 38 Å for those treated with C_{16} and $2C_{12}$ surfactants. The appearance of the CH_2 symmetric and asymmetric stretching bands on the IR spectra at around 2850 and 2920 cm^{-1}, respectively, proves successful insertion of the surfactants into clays interlayers. The values of the frequencies related to the CH_2 symmetric and asymmetric stretching bands of the surfactants in organoclays diminish and tend to those of the pure surfactants when the surfactant loading level increases. The hydrophilic and organophobic properties of the raw clays change to hydrophobic and organophilic ones upon the insertion of the surfactants into the clays interlayers. The present work shows that the organoclays synthetized by using AH and DI local clays lead to a high removal efficiency toward BPA molecules in aqueous solutions. In fact, higher values of q_m (from 88 to 127 mg/g mg/g) were achieved for BPA adsorption onto organoclays based on AH and DI clays and $n\text{-}C_{12}H_{25}(CH_3)_3N^+$, $n\text{-}C_{14}H_{29}(CH_3)_3N^+$, $n\text{-}C_{16}H_{33}(CH_3)_3N^+$ and $(n\text{-}C_{12}H_{25})_2(CH_3)_2N^+$ surfactants. The BPA removal was influenced by the adsorbent dose, the initial concentration, and the agitation time, the surfactant loading level, the pH and the temperature. The organoclays efficiencies to sorb BPA molecules from water decrease significantly when the pH value reached 10. It is also noted that when the level of surfactants is 2.0 CEC, the efficiency of the organoclays to sorb BPA is less influenced in alkaline solution than the one of those loaded at 1.0 CEC. The present work indicates that organoclays loaded at 1.0 CEC level are efficient for BPA removal from acidic solution, and those loaded at 2.0 CEC level are useful to achieve high elimination rate of BPA in alkaline solution. The results obtained from the thermodynamic and kinetic investigations show that both physisorption and chemisorption intervene in the retention of BPA species. The present work strongly sustains that organoclays prepared via solid-state intercalation are effective adsorbents as those obtained via solid-liquid reactions for BPA removal from water.

Acknowledgements

International Science Programme [ISP], Uppsala, Sweden financial support is gratefully acknowledged.

Conflict of interest

The authors declare no conflict of interest.

Author details

Issaka Garikoé and Boubié Guel*
Laboratory of Materials and Molecular Chemistry, U.F.R.—SEA/University Joseph KI-ZERBO, Ouagadougou, Burkina Faso

*Address all correspondence to: boubieguel@yahoo.fr; boubie.guel@ujkz.bf

References

[1] Nakada N, Shinohara H, Murata A, Kiri K, Managaki S, Sato N, et al. Removal of selected pharmaceuticals and personal care products (PPCPs) and endocrine-disrupting chemicals (EDCs) during sand filtration and ozonation at a municipal sewage treatment plant. Water Research. 2007;**41**:4373-4382

[2] Dong Y, Wu D, Chen X, Lin Y. Adsorption of bisphenol A from water by surfactant-modified zeolite. Journal of Colloid and Interface Science. 2010; **348**:585-590

[3] Li Y, Jin F, Wang C, Chen Y, Wang Q, Zhang W, et al. Modification of bentonite with cationic surfactant for the enhanced retention of bisphenol A from landfill leachate. Environmental Science and Pollution Research International. 2015;**22**:8618-8628

[4] Radwan M, Wielgomas B, Ziewirska ED, Radwan P, Kałużny P, Klimowska A, et al. Urinary bisphenol A levels and male fertility. American Journal of Men's Health. 2018;**12**:2144-2151

[5] Arnold SM, Clark KE, Staples CA, Klecka GM, Dimond SS, Caspers N, et al. Relevance of drinking water as a source of human exposure to bisphenol A. Journal of Exposure Science & Environmental Epidemiology. 2013;**23**: 137-144

[6] Vom Saal FS, Welshons WV. Large effects from small exposures. II: The importance of positive controls in low-dose research on bisphenol A. Environmental Research. 2006;**100**:50-76

[7] Yang Q, Gao M, Luo Z, Yang S. Enhanced removal of bisphenol A from aqueous solution by organo-montmorillonites modified with novel Gemini pyridinium surfactants containing long alkyl chain. Chemical Engineering Journal and the Biochemical Engineering Journal. 2016;**285**:27-38

[8] Park Y, Sun Z, Ayoko GA, Frost RL. Bisphenol A sorption by organo-montmorillonite: Implications for the removal of organic contaminants from water. Chemosphere. 2014;**107**:249-256

[9] Xu J, Wang L, Zhu Y. Decontamination of bisphenol A from aqueous solution by graphene adsorption. Langmuir. 2012;**28**(22): 8418-8425

[10] Zheng S, Sun PZY, Ayoko GA, Frost RL. Removal of bisphenol A from wastewater by Ca-montmorillonite modified with selected surfactants. Chemical Engineering Journal and the Biochemical Engineering Journal. 2013; **234**:416-422

[11] Liu C, Wu P, Zhu Y, Tran L. Simultaneous adsorption of Cd^{2+} and BPA on amphoteric surfactant activated montmorillonite. Chemosphere. 2016; **144**:1026-1032

[12] Cao F, Bai P, Li H, Ma Y, Deng X, Zhao C. Preparation of polyethersulfone-organophilic montmorillonite hybrid particles for the removal of bisphenol A. Journal of Hazardous Materials. 2009; **162**:791-798

[13] Murugananthan M, Yoshihara S, Rakuma T, Shirakashi T. Mineralization of bisphenol A (BPA) by anodic oxidation with boron-doped diamond (BDD) electrode. Journal of Hazardous Materials. 2008;**154**:213-220

[14] Wang ZM, Ooga H, Hirotsu T, Wang WL, Wu QY, Hu HY. Matrix-enhanced adsorption removal of trace BPA by controlling the interlayer hydrophobic

environment of montmorillonite. Applied Clay Science. 2015;**104**:81-87

[15] Bhatnagar A, Anastopoulos I. Adsorptive removal of bisphenol A (BPA) from aqueous solution: A review. Chemosphere. 2017;**168**:885-902

[16] Rathnayake SI, Xi Y, Frost RL, Ayoko GA. Environmental applications of inorganic-organic clays for recalcitrant organic pollutants removal Bisphenol A. Journal of Colloid and Interface Science. 2016;**470**:183-195

[17] Liu S, Wu P, Chen M, Yu L, Kang C, Zhu N, et al. Amphoteric modified vermiculites as adsorbents for enhancing removal of organic pollutants: Bisphenol A and tetrabromo bisphenol A. Environmental Pollution. 2017;**228**:277-286

[18] Cheng C, Ma L, Ren J, Li L, Zhang G, Yang Q, et al. Preparation of polyethersulfone-modified sepiolite hybrid particles for the removal of environmental toxins. Chemical Engineering Journal and the Biochemical Engineering Journal. 2011;**171**:1132-1142

[19] Zhao Z, Fu D, Ma Q. Adsorption characteristics of bisphenol A from aqueous solution onto HDTMAB-modified palygorskite. Separation Science and Technology. 2014;**49**:81-89

[20] Guo W, Hu W, Pan J, Zhou H, Guan W, Wang X, et al. Selective adsorption and separation of BPA from aqueous solution using novel molecularly imprinted polymers based on kaolinite/Fe_3O_4 composites. Chemical Engineering Journal and the Biochemical Engineering Journal. 2011;**171**:603-611

[21] Dehghani MH, Ghadermazi M, Bhatnagar A, Sadighara P, Jahed-Khaniki G, Heibati B, et al. Adsorptive removal of endocrine disrupting bisphenol A from aqueous solution using chitosan. Journal of Environmental Chemical Engineering. 2016;**3**:2647-2655

[22] Li J, Zhan Y, Lin J, Jiang A, Xi W. Removal of bisphenol A from aqueous solution using cetylpyridinium bromide (CPB)-modified natural zeolites as adsorbents. Environment and Earth Science. 2014;**72**(10):3969-3980

[23] Goyal N, Bulasara VK, Barman S. Surface modification of synthesized nanozeolite NaX with TEAOH for removal of bisphenol A. Chemical Engineering Communications. 2016;**203**(10):1374-1384

[24] Betega de Paiva L, Morales AR, Valenzuela DFR. Organoclays: Properties, preparation and applications. Applied Clay Science. 2008;**42**:8-24

[25] Mishra AK, Allauddin S, Narayan R, Aminabhavi TM, Raju KVSN. Characterization of surface-modified montmorillonite nanocomposites. Ceramics International. 2012;**38**:929-934

[26] Wang H, Zhang H, Jiang JQ, Ma X. Adsorption of bisphenol A onto cationic-modified zeolite. Desalination and Water Treatment. 2016;**57**:26299-26306

[27] Garikoé I, Sorgho B, Guel B, Persson I. Solid-state synthesis and physicochemical characterization of modified smectites using natural clays from Burkina Faso. Bulletin of the Chemical Society of Ethiopia. 2021;**35**:43-59

[28] Park Y. Synthesis, characterisation and application of organic surfactants modified clays for water purification. PhD diss., Queensland University of Technology. 2013

[29] Kamitori S, Sumimoto Y, Vongbupnimit K, Noguchi K, Okuyama K. Molecular and crystal structures of

dodecyltrimethylammonium bromide and its complex with P-phenyl-phenol. Molecular Crystals and Liquid Crystals. 1997;**300**:31-43

[30] Lagaly G, Ogawa M, Dékány I. Handbook of Clay Science, Chap 7.3: Clay mineral organic interactions. Vol. 1. 1st ed. Developments in Clay Science. Amsterdam: Elsevier; 2006. pp. 309-330

[31] Madejová J. FTIR techniques in clay mineral studies. Vibrational Spectroscopy. 2003;**31**:1-10

[32] Madejová J, Bujdak J, Janek M, Komadel P. Comparative FT-IR study of structural modifications during acid treatment of dioctahedral smectites and hectorite. Spectrochimica Acta. Part A, Molecular and Biomolecular Spectroscopy. 1998;**54**:1397-1406

[33] Cubuk O, Caglar B, Topcu C, Coldur F, Sarp G, Tabak A, et al. Structural characterization of hexadecyltrimethylammonium-smectite composites and their potentiometric electrode applications. Applied Surface Science. 2015;**338**:99-112

[34] Silva MR, Beja AM, Paixao JA. Pseudosymmetry in tetradecyltrimethyl-ammonium bromide. Acta Crystallographica. 2003;**E59**:01151-01152

[35] Campanelli AR, Scaramuzza L. Hexadecyltrimethylammonium bromide. Acta Crystallographica. 1986; **C42**:1380-1383

[36] Li Z, Jiang W-T, Hong H. An FTIR investigation of hexadecyltrimethylammonium intercalation into rectorite. Spectrochimica Acta. Part A, Molecular and Biomolecular Spectroscopy. 2008;**71**:1525-1534

[37] He H, Ma Y, Zhu J, Yuan P, Qing Y. Organoclays prepared from montmorillonites with different cation exchange capacity and surfactant configuration. Applied Clay Science. 2010;**48**:67-72

[38] Caglar B, Topcu C, Coldur F, Sarp G, Caglar S, Tabak A, et al. Structural, thermal, morphological and surface charge properties of dodecyltrimethylammonium-smectite composites. Journal of Molecular Structure. 2016;**1105**:70-79

[39] Garikoé I, Sorgho B, Yaméogo A, Guel B, Andala D. Removal of bisphenol A by adsorption on organically modified clays from Burkina Faso. Bioremediation Journal. 2021;**25**:2-47

[40] Liu Y, Gao M, Gu Z, Luo Z, Ye Y, Lu L. Comparison between the removal of phenol and catechol by modified montmorillonite with two novel hydroxyl containing Gemini surfactants. Journal of Hazardous Materials. 2014; **267**:71-80

[41] Sarkar B, Megharaj M, Xi Y, Naidu R. Surface charge characteristics of organo-palygorskites and adsorption of p-nitrophenol in flow-through reactor system. Chemical Engineering Journal and the Biochemical Engineering Journal. 2012;**185-186**:35-43

[42] Özcan AS, Özcan A. Adsorption of acid dyes from aqueous solutions onto acid-activated bentonite. Journal of Colloid and Interface Science. 2004;**276**: 39-46

[43] Liu Y. Is the free energy change of adsorption correctly calculated? Journal of Chemical & Engineering Data. 2009; **54**:1981-1985

[44] Ding L, Zou B, Gao W, Liu Q, Wang Z, Guo Y, et al. Adsorption of

Rhodamine-B from aqueous solution using treated rice husk-based activated carbon. Colloids and Surfaces A: Physicochemical and Engineering Aspects. 2014;**446**:1-7

[45] Tong DS, Zhou CH, Lu Y, Yu H, Zhang GF, Yu WH. Adsorption of acid red G dye on octadecyltrimethylammonium montmorillonite. Applied Clay Science. 2010;**50**:427-431

[46] Vithanage M, Mayakaduwa SS, Herath I, Ok YS, Mohan D. Kinetics, thermodynamics and mechanistic studies of carbofuran removal using biochars from tea waste and rice husks. Chemosphere. 2016;**150**:781-789

[47] Kubono A, Yuasa N, Shao H-L, Umemoto S, Okui N. Adsorption characteristics of organic long chain molecules during physical vapor deposition. Applied Clay Science. 2002;**193**:195-203

[48] Vuković GD, Marinković AD, Škapin SD, Ristić M, Aleksić R, Perić-Grujić AA, et al. Removal of lead from water by amino modified multiwalled carbon nanotubes. Chemical Engineer. 2011;**173**:855-865

[49] Björklund K, Li LY. Adsorption of organic stormwater pollutants onto activated carbon from sewage sludge. Journal of Environmental Management. 2017;**197**:490-497

[50] Yousef RI, El-Eswed B, Al-Muhtaseb AH. Adsorption characteristics of natural zeolites as solid adsorbents for phenol removal from aqueous solutions: Kinetics, mechanism, and thermodynamics studies. Chemical Engineering Journal and the Biochemical Engineering Journal. 2011;**171**:1143-1149

[51] Errais E, Duplay J, Darragi F, M'Rabet I, Aubert A, Huber F, et al. Efficient anionic dye adsorption on natural untreated clay: Kinetic study and thermodynamic parameters. Desalination. 2011;**275**:74-81

[52] Yu Y, Zhuang YY, Wang ZH, Qiu MQ. Adsorption of water-soluble dyes onto modified resin. Chemosphere. 2004;**54**:425-430

[53] Liu QS, Zheng T, Wang P, Jiang JP, Li N. Adsorption isotherm, kinetic and mechanism studies of some substituted phenols on activated carbon fibers. Chemical Engineering Journal and the Biochemical Engineering Journal. 2010; **157**:348-356

[54] Genç N, Kılıçoğlu Ö, Narci AO. Removal of Bisphenol A aqueous solution using surfactant-modified natural zeolite: Taguchi's experimental design, adsorption kinetic, equilibrium and thermodynamic study. Environmental Technology. 2017;**38**: 424-432

Chapter 5

Chitosan-Based Nanocomposites for Biological Applications

Serap Yalcin and Nevin Cankaya

Abstract

Chitosan is an important natural cationic polymer. Chitosan is produced as a deacetylated form of chitin, and its excellent biocompatible, biodegradable, nontoxic, natural chemical, and thermal stability properties have led to its common use in especially biomedical applications. The combination of nanomaterials and chitosan has been considered an excellent approach to overcoming the handicaps associated with biopolymer. The chitosan-based nanocomposites are potentially efficient in a number of areas including medical fields. Chitosan is biodegradable, biocompatible, basic, nontoxic, and also approved by GRAS (Generally recognized as safe by the United States Food and Drug Administration [US FDA]). Chitosan-based nanocomposites have different applications in drug delivery including ocular, per-oral, pulmonary, nasal mucosal, gene, buccal drug, vaccine, vaginal, and cancer therapy. Chitosan has low toxicity in both in vitro and in vivo models. *In this* chapter, we discussed the preparation techniques and various forms of chitosan materials in biomedical applications. In addition, this chapter explores recent research on chitosan-based nanocomposites for medical studies.

Keywords: chitosan, chitosan-based nanocomposites, clay, biomedical applications

1. Introduction

Chitin is the most abundant polysaccharide in nature after cellulose and is commercially produced from the waste shells of lobster, shrimp, crabs, etc. obtained. Chitosan is obtained by removing the acetyl group of chitin from the structure in a basic medium. Chitosan is a copolymer formed by connecting N-acetyl-D-glucose amine and D-glucose amine with β-1,4 glycosides bonds (**Figure 1**). Since chitosan contains free amino groups, it is a neutral polysaccharide at neutral and basic pH and is insoluble in water. However, at acidic pH, amino groups are soluble because they are protonated. Its solubility depends on the distribution of free amino and N-acetyl groups and is readily soluble in dilute 1–3% acetic acid [1]. Chitosan does not cause allergic reactions and is biocompatible with living tissues. Amino sugars are gradually broken down into harmless products in the body and are completely absorbed by the human body, and they can be easily removed from the organism without causing local side effects in the body [2]. Chitosan is a bioactive ingredient with numerous properties such as antitumor, immune enhancer, antifungal, antimicrobial, antioxidant, and

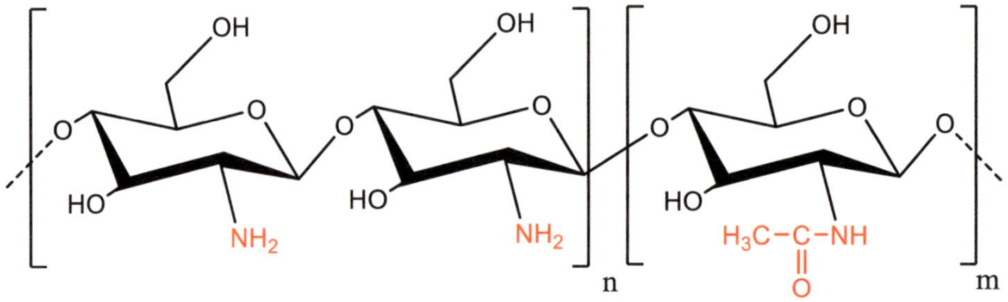

Figure 1.
Chemical structure of chitosan.

wound healing. In addition to these features, it also has features such as being biodegradable, biocompatible, low cost, and non-antigenic [3].

Chitosan also has good adhesion and coagulation properties. The presence of primary amine groups and primary and secondary hydroxyl groups in chitosan makes it very useful in biological applications. Compared to other biopolymers, chitosan is positively charged and has a structure that can adhere to the mucosa [4]. Chitosan is widely used in many industrial applications due to the following properties [1, 5].

- It is renewable and abundant.
- It has Nontoxic, biocompatible, and biodegradable properties.
- It has bioeffects such as antacid, antitumor, antioxidant, and antibacterial.
- No harmful in/organic solvents are required for dissolution.
- It is a cationic biopolymer and interacts easily with negatively charged surfaces.
- It has strong electrostatic properties and shows suitable bioadhesive properties.
- Electrostatically strong and exhibits favorable bioadhesive properties.

2. Properties of chitosan

The molecular structure of chitosan, which is obtained by partial deacetylation of chitin and contains β-(1,4)-D-glucosamine and N-acetyl-D-glucosamine units, is shown in **Figure 1** [5].

2.1 Physical and chemical properties of chitosan

There are some parameters that affect the characteristic properties of chitosan, such as molecular weight, degree of deacetylation, and solubility. These parameters vary according to the conditions applied during the production of chitosan [6]. The process of removing acetyl groups in the chitin polysaccharide chain is called deacetylation. Chitin, which usually has a degree of deacetylation greater than 50%, is considered to be chitosan [7, 8]. The molecular weight of chitosan varies depending on the conditions applied during the deacetylation process and the source from which it is obtained. For example, exposure of chitosan to oxygen and high temperature causes its molecular weight to decrease [9].

Some parameters affecting the physical and chemical properties of chitosan are as follows:

2.1.1 Resolution

Thanks to its cationic structure, chitosan can be easily dissolved in some solutions in pH <6 environments. Organic acids such as acetic acid, formic acid, and lactic acid are generally used to dissolve chitosan. On the other hand, the solubility of chitosan in inorganic acids is quite low [9]. Some mineral acids such as hydrochloric acid and nitric acid can also be dissolved, but it is not suitable to use phosphoric acid and sulfuric acid as solvents [10, 11]. Among them, the most widely used is acetic acid. The solubility of chitosan mostly depends on the degree of N-acetylation (or degree of deacetylation), distribution of acetyl groups, deacetylation time, pH, ionic strength [12], temperature, particle size, pre-treatments before isolation, base concentration, and chitin/base solution ratio [13].

The solubility problem of chitosan is a disadvantage and must be used by dissolving in acid. In order to improve this, water-soluble chitosan derivatives can be synthesized, thanks to the reactivity of the primary amine and primary and secondary hydroxyl groups in its structure. Modification of chitosan increases its solubility and stability, making it versatile as a biopolymer [14].

2.1.2 Deacetylation degree (DD)

The degree of deacetylation of chitosan plays an important role in its solubility and solution properties [15]. DD and molecular weight have the effect of changing the physicochemical and biological properties of chitosan. Studies have been carried out with different chitosan samples with similar molecular weights but varying DD between 70% and 95%, and these parameters were found to be related to physicochemical properties. Chitosan samples with high DD have more crystalline regions than those with low DD. As the DD increases, the elasticity and tensile strength also increase [16]. In terms of biological properties, chitosan samples with DD exceeding 90% were examined. It has been observed that chitosan samples with high DD play a role in the regeneration of nerve cells and resemble cells in the peripheral nervous system. As a result, the DD of chitosan is a very important parameter in terms of physicochemical and biological properties [17].

2.1.3 Molecular weight (Mw)

The Mw of chitosan varies depending on its source and deacetylation conditions (temperature, time, and base concentration). The dissolved oxygen in the solution medium reduces the stability of chitosan, causing it to decompose, and the Mw of chitosan decreases. Also, high temperature ($\geq$280°C) breaks the polymer chains of chitosan and lowers its Mw [9]. In the literature, it is stated that high molecular weight polymers slow down the release rate of drugs [18, 19]. The Mw, viscosity, polarity, solubility, and thermal properties of the drug carrier material matrix significantly affect the release mechanism [19].

2.1.4 Viscosity

Many factors affect the viscosity of chitosan, such as the degree of deacetylation, molecular weight, ionic strength, pH, and temperature [9]. Viscosity increases with an increasing degree of deacetylation. Chitosan with a high and low degree of deacetylation has different conformations in an aqueous solution. Chitosan with a high degree of deacetylation has an expanded conformation with more flexible chains due to charge repulsion in the molecule. However, chitosan with low DD is rod-like or coil-shaped due to the low charge density in the polymer chains. The viscosity of chitosan is also affected by concentration and temperature. If the concentration of the medium increases or the temperature decreases, the viscosity of chitosan also increases [13].

The viscosity of chitosan is also highly related to its Mw. The viscosity of chitosan with a high Mw is higher than that of a low Mw one. Many studies show that physical and chemical processes affect viscosity. Processes such as increasing the grinding time, heating, autoclaving, ultrasonic, and ozonation reduce the viscosity [20]. It was stated that the viscosity decreased with increasing deproteinization and demineralization time [9, 20].

2.1.5 Color

The pigment in the shellfish structure forms a complex with chitin. The color of powdered chitosan can vary from light yellow to white. When obtaining chitosan from chitin, decolorization can be done by extraction with acetone followed by interaction with 0.3% NaOCl. Chemicals such as $KMnO_4$, $NaHSO_3$, $Na_2S_2O_4$, and H_2O_2 are also used for color removal [9].

The amino and hydroxyl functional groups in chitosan allow it to form stable covalent bonds with other materials. It can carry out esterification and etherification reactions with alcohol groups, while amino groups on D-glucosamine can be quaternized or react with aldehyde functions [21]. Thanks to its amino groups, chitosan can form complexes with metal ions, and thus it can be used in the treatment of wastewater and recovery of heavy metals [22, 23]. The complexing feature of chitosan also enables it to be used in the purification of beverages (juices, wine, etc.) [22].

Chitin has a stable structure and is insoluble in water, alcohol, dilute acid, and base solutions, and its chemical reactivity is quite low. Due to these features, it is not widely used in industrial applications [23, 24]. Today, chitosan is used in numerous fields such as food, agriculture, cosmetics, textile, medicine, and pharmacy sectors. Thanks to the mentioned properties of chitosan, its field of study can be further increased by its modifications with various functional groups in the future.

3. Biomedical applications chitosan-based nanocomposites

3.1 Controlled drug delivery systems

Before controlled release systems were developed, there were many systems with long-acting, different drug releases, and different names. Modified release systems, which differ from basic drug delivery methods, fall into two groups:

1. Delayed-release systems

2. Extended-release systems:

- Controlled release systems
- Sustained-release/prolonged systems

The main systems defined in the modified release systems classification and differing from each other in active substance releases are as follows [25, 26]:

Repeat Action Systems: There is more than one dose of the active substance in the applied dosage, and these doses are released at certain time intervals.

Delayed-Release Systems: The release of the active substance from the system takes place in a certain region. It is used for enteric-coated tablets, generally.

Sustained Release Systems: These systems can maintain the plasma and tissue levels of the active substance for a longer period of time than conventional release systems. However, since the system is affected by environmental conditions, it is difficult to determine the release mechanism in advance. In general, its velocity is compatible with first-order kinetics.

Controlled Release Systems: These systems exhibit different behavior from sustained-release systems in that release rates can be planned in advance and can realize drug release with zero-order kinetics [25, 27].

Controlled release is a constantly evolving topic with applications in many different fields such as medicine, pharmacy [28], chemistry [29], environment, agriculture [30], and veterinary medicine [31]. Thanks to the controlled release systems, the pollution caused by the traditional application methods in the soil can be prevented by using low amounts in agriculture (agricultural pesticides such as fertilizers and insecticides) and applications related to environmental protection [30]. It is used in controlled release applications of parasitic drugs, hormones, vaccines, antibiotics, substances that increase milk yield, and birth control drugs in veterinary medicine [31]. In chemical applications, it is also used for the controlled release of expensive and waste-producing materials, thus ensuring economic and continuous production. Controlled release practices in medicine and pharmacy have emerged in order to better control drug administration, facilitate the treatment of the patient, and increase the quality of life of the patient [32].

Controlled drug release is a method in which the active substance is designed to be released from a system at the desired time, at a determined rate, and in the required quantities [33]. The interest in controlled drug delivery systems is increasing day by day because developing a new drug takes a long time and is economically burdensome. Thanks to controlled drug release systems, the drug dose used decreases, the dosage range increases, and the side effects of the drug are largely eliminated [32]. After the active substance mixes into the blood, its level blood should remain within the plasma range where it is effective. In classical release systems, the drug level fluctuates between the maximum and minimum values, and when the drug life is over, it is necessary to take the drug in high doses again. In controlled drug release systems, the drug can be released at the optimum plasma concentration, in a stable manner, and over a long period of time [25, 34].

In order for the drug to be effective, it must first form the dosage form that carries the active substance, then mix it into the blood safely and effectively, disperse into the tissues, and show its effect in the target area. The dosage should be maintained in a range above the effective amount and below the toxic amount after mixing with the blood. Each dose of the drug taken reaches a peak according to its specific half-life in the blood, then decreases below the effective amount, and is finally eliminated from the body. In conventional drug systems, it is not possible for the drug to select its place of action or to mix into the blood in a controlled manner [25, 35]. Controlled drug release application, it is aimed to show the effectiveness of the drug according to a pre-planned process in the body and to perform treatment at longer intervals, with low doses, without side effects. Thus, the life of the drug in its circulation in the body is prolonged, the absorption is accelerated and its target ability to the effect site is ensured [36].

The conventional and controlled exchange of the drug in the blood is shown schematically in **Figure 2**.

Drug delivery systems provide predetermined and reproducible controlled drug release for long-term treatment locally or systemically at specified time intervals. In the traditional method, the drug is given at once and in high doses, and the dose is repeated after a few hours or a day. This method is not economical and has side effects [37, 38]. The purpose of controlled drug release systems is to improve the performance of drug therapy. This mechanism enhances the therapeutic activity and reduces side effects by reducing the toxicity caused by overdose during treatment. If the drug concentration level is not stable, the drug falls below the normal level or rises above the toxic level. This may cause undesirable side effects in the patient. The controlled drug release system maintains a constant level of drug concentration in the blood plasma [39, 40].

Controlled drug delivery systems are produced by combining a polymer with a drug or other active agent. Polymers used in drug delivery systems should be nontoxic and non-allergenic, high purity and reusable, biodegradable in vivo, and structures formed after the degradation process should be usable in metabolism [38]. These polymers used as biomaterials can be natural or synthetic. Despite the advantage that synthetic polymers can be synthesized at any time, biopolymers have advantages such as being easily obtained from nature, cheap, and chemically modified [1, 5]. Chitosan has a very important role in medical applications due to its nontoxicity, biocompatibility, biodegradability, mucoadhesion, and low allergenic. Its high biocompatibility and biodegradable properties are its most important advantages for drug delivery systems [38, 39].

3.2 Advantages and disadvantages of controlled drug delivery systems

Controlled release systems are used successfully in the treatment of many diseases today. In the coming years, with the use of controlled-release drugs (especially tablets and capsules), the effectiveness and safety of the treatment will increase. The

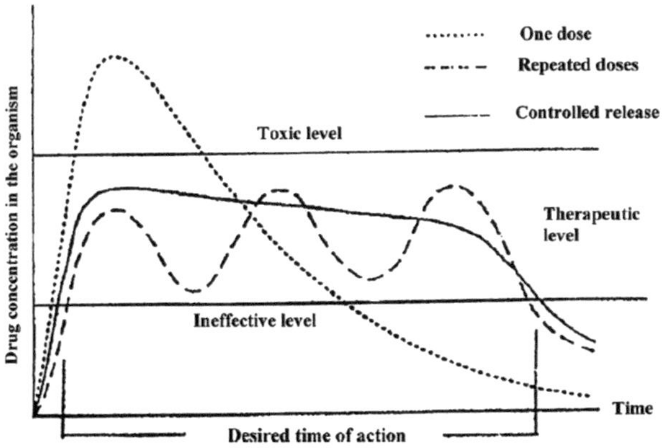

Figure 2.
A schematic drawing of drug release [12].

advantages and disadvantages of controlled release systems can be summarized as follows [25, 32–42]. Advantages of controlled-release systems:

- The plasma level of the drug can be kept at the desired value for the specified time.
- Thanks to infrequent drug intake, drug compliance increases.
- The total drug dose loaded into the system is less than conventional dosage forms.
- Eliminating or reducing the toxic and side effects of the drug.
- Preventing the degradation of drugs with short in vivo half-lives and prolonging their half-life.
- Controlled release systems can be targeted to the desired region, organ, and tissue.
- It provides convenience to patients, their relatives, and caregivers, and drug use can be controlled, especially in underdeveloped regions where medical control cannot be fully achieved.

Despite all these benefits, there are still unsolved problems with controlled release systems. Main disadvantages:

- The polymer itself or the decomposed products may exhibit toxic effects or biocompatibility.
- After the system is used, the drug release cannot be stopped at any time.
- Reliability of the system is weak in case of physical changes that may occur during production, storage, and distribution (capillary cracks, etc.).
- It may not be possible to prepare every drug with this system, and there is no single technology that can be applied to drugs.
- This system may increase the cost of drugs.

In some of these systems, the cost may be greater than in conventional dosage forms. However, well-designed controlled release systems are preferred because they reduce overall health expenditures when evaluated in terms of cost-effectiveness. Controlled release systems of cancer, hormone, enzyme, antibiotic, antirheumatic drugs, and some existing drugs produced in biotechnology or classical technologies have been prepared. Thanks to the developments in molecular medicine in recent years, controlled release systems have also given successful results in the use of protein and peptide drugs.

3.3 Chitosan-based clay-containing material applications in drug delivery systems

For controlled drug release systems, some carrier materials such as natural or synthetic polymers, metals, ceramics, oils, antibodies, magnetic components, and carbon are used in the world today. The most widely used of these materials are polymers, and some of these studies are in the stage of animal trials. Biocompatibility and biodegradability properties are sought in polymers used in drug delivery systems. Biodegradable polymers are polymers that can be enzymatically, microbiologically, or hydrolytically degraded in a physiological environment. The degradation products of biodegradable polymers must also be nontoxic. Biopolymers, modified natural polymers, and synthetic polymers are frequently used in the preparation of drug delivery systems. Along with this polymer, chemicals such as calcium hydroxyapatite, clay, organic-containing clay derivatives, magnetite and maghemite nanoparticles (for magnetic control), and graphene oxide are also used. In the following paragraphs, these subjects will be explained and studies on chitosan will be emphasized.

In recent years, studies have been carried out on the preparation and application of biodegradable polymer nanocomposites as controlled drug delivery systems [38–40]. Clay types such as montmorillonite (MMT), halloysite, organoclay, etc. are used in the preparation of these nanocomposites. MMT is a widely used material for controlled drug release as it retains the drug, thanks to its high cation exchange capacity. Clay minerals are natural cation exchangers and bind to the drug in solution by electrostatic attraction. Depending on the cation exchange capacity of the clay, cationic of the drug, release medium, and pH decide the release kinetics of the drug. It is possible to have different interactions such as hydrophobic, hydrogen bonding, ligand exchange, and water bridge, independent of electrostatic forces. These properties of clay encourage the use of clay for sustainable drug release [43]. However, the ability of clay particles to adsorb negative charged or neutral drugs is low. This limits the applications of negatively charged or neutral drugs. When preparing drug delivery systems, this disadvantage of clay should be taken into account [43].

Yuan et al. observed that the swelling rates of chitosan-clay biocomposites that they synthesized for drug release were lower than clay alone and chitosan alone. However, they stated that the drug release from the chitosan-clay biocomposite was more than that of clay and less than chitosan. They stated that with the addition of clay to chitosan, the interaction between the negatively charged groups in the clay and the positively charged NH_3^+ groups in the chitosan creates strong crosslinks. They stated that this situation affects the swelling behavior of the composite and, therefore, affects the diffusion of the drug in the structure [43]. Hua et al. studied the release of the drug ofloxacin from chitosan-MMT and chitosan-only hydrogel. They prepared different ratios of drug and clay with the same amount of chitosan, and they observed that the hydrogel containing the highest clay had the highest drug retention efficiency. With its high surface area, MMT adsorbed the drug not only on the inner and outer surfaces but also between the clay layers. With the addition of MMT, the drug release decreased and the dispersion viscosity of the drug increased. In the swelling test, the least swelling value was observed in chitosan, and the maximum swelling value was observed in the hydrogel with the highest MMT content. The amount and rate of drug release decreased as the amount of MMT increased. In the XRD analysis, the presence of the drug was observed in the medicated composite and chitosan, while the drug was lost in the hydrogel with high clay content. This indicates that drug molecules are lost at the polymeric level or get into the clay layers [44]. Cheikh et al. synthesized nanocomposite by intercalation method using chitosan and purified beidellite. In this study, they selected diclofenac sodium as a model drug and examined its release. According to the results of the analysis, the drug was both distributed between the layers of the clay and detected on the surface of the nanocomposite. They reported that with the nanocomposites they obtained, they reached 60% cumulative drug release in 8 hours at pH 6.8 [45]. In another study, Sharma et al. synthesized silymarin-loaded chitosan-MMT microbeads by ionotropic gelation method for the potential treatment of gastric ulcer [46]. Depan et al. grafted lactic acid onto chitosan and obtained the chitosan-g-lactic acid/Na^{+}MMT bionanocomposite. They characterized the structure of the composite by FTIR, XRD, SEM, and ^{1}H-NMR, and thought that this bionanocomposite could be used in controlled drug release and tissue engineering. For this purpose, they used the synthesized bionanocomposite for the transport of the drug sodium ibuprofen [47]. Sahoo et al. synthesized Chitosan/polycaprolactone/Cloisite30B bionanocomposites using an 80:20 ratio of chitosan: polycaprolactone and Cloisite30B (organoMMT) at 1, 2.5 and 5% by mass and characterized their structures by FT-IR, XRD, and SEM. They reported that the synthesized biocomposite could be used in the controlled release of the drug doxycycline [48]. Parida et al. synthesized the Chitosan/Polyvinyl Alcohol/Cloisite30B bionanocomposite using Cloisite30B at different rates and reported that the composite could be used in controlled drug release [49]. In another study, Nanda et al. synthesized (by solvent removal method) Chitosan/Polylactic Acid/Cloisite30B bionanocomposites using different ratios of chitosan, polylactic acid, and Cloisite30B and characterized their structures with FT-IR, SEM, and XRD. They stated that the drug release properties of these biocomposites, which they used in the controlled release of paclitaxel anticancer drugs, were sensitive to pH and the amount of drug loaded [50]. Cojocariu et al. synthesized the Chitosan/Cloisite15A (organoMMT) bionanocomposite using different amounts of clay. They stated that the bionanocomposite has an intercalated structure by XRD and SEM analysis. They reported that the bionanocomposite containing 9% by mass of Cloisite15A delayed the controlled drug release [51].

It is known that hydrogels as new drug delivery systems have been used extensively in controlled drug release in recent years. Hydrogels provide control of drug release by showing swelling-shrinking behavior at different rates in different environments (temperature, pH, humidity, electric current, magnetic field, light, etc.) according to the cross-linker ratios in their content. Similarly, in nanoparticle synthesis, hydrogels are used for controlled size adjustment and stabilization of the produced nanometal particles [52]. Wang et al. synthesized a series of polymeric materials containing different amounts of attapulgite clay from the material they prepared with pH-sensitive chitosan, acrylic acid, attapulgite, and sodium alginate. Diclofenac sodium active drug substance was used to examine the controlled drug loading and release kinetics of the prepared materials. They reported that the material released 100% drug in basic medium and 3.76% in acidic medium [53]. Dinu et al. prepared chitosan/clinoptilolite biocomposite cryogels containing clinoptilolite clay in different proportions and investigated their drug release properties. SEM photographs revealed that the pores became smaller and the pore walls thickened as the clay content in the biocomposite increased. They reported that the swelling behavior of the cryogels they obtained and the drug release properties showed parallelism [54]. Luo and Mills used halloysite clay to strengthen chitosan hydrogels and prepared a biocompatible and biodegradable drug delivery system. They loaded gentamicin on both halloysite clay and chitosan-halloysite hydrogel. They observed that drug-loaded chitosan-halloysite hydrogels released drugs more slowly than drug-loaded halloysite hydrogels. They stated that as the rate of chitosan increased, long and effective drug release occurred over time [55]. Hua et al. synthesized chitosan/ofloxacin/MMT nanocomposite hydrogels using ofloxacin, a synthetic antibiotic, using sodium tripolyphosphate as the anionic cross-linker. In the drug delivery system, while the chitosan beads deteriorated within 3 hours at pH=1.2, it was determined as a result of the analysis that the synthesized nanocomposite hydrogel degraded in 12 hours [55]. Ma et al. synthesized pH and temperature-sensitive carboxymethyl chitosan-g-poly(N, N-dimethyl acrylamide) polymers. They used vitamin B2 as a model drug and aimed to achieve intestinal-targeted controlled release of the developed drug-loaded hydrogels. Hydrogel beads were prepared by Ca^{2+} ionic crosslinking in an acidic solution and a double crosslinked network structure was obtained. The morphological features of the obtained products were also characterized. It was determined that the synthesized hydrogels performed an effective controlled release under gastrointestinal system conditions [56]. Aycan and Alemdar used hydroxyapatite-natural bone powder to increase the thermal and mechanical strength of chitosan hydrogels. They investigated the controlled and effective release of the drug under different physical conditions of the stomach and intestinal environment by loading the active ingredient of amoxicillin, which is used in the treatment of gastric ulcers, into hydrogels [57]. Yücel et al. prepared chitosan nanoparticles with quercetin, which is one of the polyphenolic compounds with antioxidant properties, and conducted in vitro and release studies. They observed that the quercetin release of nanoparticles could be sustained for 24 hours [58]. İnal investigated the release of indomethacin drug by synthesizing chitosan/κ-carrageenan/chitosan trilayer microspheres. Controlled release studies were performed in pH 1.2 and 7.4 buffers and characterized by FTIR, SEM, and DSC. The drug entrapment efficiency of microspheres and the equilibrium swelling degree were determined by the particle size and controlled release data. He stated that the newly obtained system is a suitable controlled release system for drugs that cause stomach irritation [4]. Ulu A. synthesized allantoin-loaded chitosan nanoparticles and investigated the effect of chitosan molecular weights (low, medium, and high) on drug release. He proved that allantoin-loaded nanoparticles are affected by the molecular weight of chitosan by morphology, size, zeta potential, and loading efficiency methods. From the in vitro release results, it was observed that nanoparticles synthesized with chitosan with the lowest molecular weight were more effective in drug release [59].

Hospital infections significantly affect the success of implant materials placed in the body. Önder S. conducted a controlled release study of gentamicin, an antibiotic type, on a chitosan/titanium system that can be used to prevent infections. He first immobilized the antibiotic-laden chitosan onto titanium surfaces, and then exposed it to drying conditions (air and freeze-drying). It was observed that the release of antibiotics was higher on the freeze-drying surfaces. From the cell proliferation tests, it was observed that bone cells proliferated more in the chitosan/titanium system compared to the flat chitosan-coated surfaces [60]. In order to prevent infections related to implants and increase tissue interaction, studies are also carried out on the functionalization of implant surfaces and their drug release. Erşan and Önder S., in another of their studies, synthesized and characterized chitosan microspheres that can be used as bone filling material and drug delivery system, and determined the performance of microspheres in vitro. Microspheres were produced by the emulsion cross-linking method and used antibiotic ciprofloxacin as a drug. They tested the bioactivity of the microspheres in artificial body fluid and found that chitosan microspheres from the bioactivity tests had the potential to increase osteointegration. They stated that these spheres could be used as a filling material that can release biomolecules locally in bone tissue damage and can be used in surface modification of implant materials [38].

In drug delivery systems, functionalized, superparamagnetic magnetite (Fe_3O_4), and maghemite (ɣ-Fe_2O_3) nanoparticles enable the drug to reach the desired target cells with the effect of an externally applied magnetic field. Thus, the side effects of the drug are minimized. The surfaces of these nanoparticles, which are used as carrier systems, are usually functionalized with drugs, proteins, and genetic materials; and with these particles, the therapeutic agent is released at the targeted site [61]. Magnetite and maghemite nanoparticles are of great importance for biomedical applications because they are biocompatible and show low levels of toxicity [62]. Most of the applications in this field are biological investigations used to the orientation of the nanoparticle with the help of an externally applied magnetic field. Such materials, prepared by embedding ferromagnetic particles into the gel, are placed inside the body, the magnetic field is activated by a device used to provide a magnetic field, and the drug in the gel begins to be released [63]. Mahdavinia et al. synthesized magnetic and pH-sensitive hydrogel spheres obtained from carrageenan and carboxymethyl chitosan for a drug delivery system. Magnetic Fe_3O_4 nanoparticles were added to the biopolymer mixture by in situ polymerization method. The structural properties of the hydrogel spheres were characterized by TEM, SEM, XRD, and VSM techniques. The pH-dependent swelling behavior of hydrogels was investigated in various buffer solutions, and the swelling capacity of magnetic hydrogels was affected by the incorporation of magnetite nanoparticles into the carrageenan/chitosan complexes. The water absorption capacity of hydrogels decreased with increasing magnetite content. In the study, methotrexate, an anticancer drug was loaded into hydrogels as a model drug, and its release profiles were investigated at pH 7.4 and 5.3. Methotrexate encapsulation efficiency increased with increasing magnetite and chitosan content [64]. Long et al. synthesized chitosan/carrageenan/Fe_3O_4 nanoparticles, adsorbed bovine serum albumin into them, and studied the release behavior of protein from the nanoparticles [65]. Wang et al. prepared magnetic chitosan-5-fluorouracil nanoparticles for the target drug delivery system. The results showed that the loading capacity was 13.4%, and the percentage of release in phosphate buffer solution (pH=7.2) was 68% at 30 hours [66].

4. Other applications of Chitosan-based nanocomposites

Significant attention has been paid to many kinds of biomedical fields because of Chitosan-based nanomaterials due to their special chemical properties, including desired biodegradability, compatibility, and nontoxicity. Chitosan is a convenient biomaterial to construct extracellular tissue matrixes in tissue engineering [67]. Chitosan can be used as a carrier for drug delivery and also for different types of therapeutic molecules such as genes, drugs, and proteins [68]. It is greatly used as a carrier in delivering active agents and drugs [69], in gene and cancer therapy [70], and also in biosensor monitoring and bioimaging [71, 72]. Chitosan behaves like an anti-plaque agent and can intervene with all microorganisms while performing antibacterial activities in dentistry [73]. Chitosan is more generally used in wound dressing as an antimicrobial and antifungal agent because of its perfect tissue adhesive features (**Figure 3**).

4.1 Wound healing applications of Chitosan-based nanocomposites

Polymer nanocomposites are described as sophisticated materials, which carry nanoparticles. They can also be presented as core-shell nanoparticles. Chitosan produced an amino group that can be operationalized further to be reconciled for a great variety of applications. Antimicrobial chitosan nanocomposites are also attractive in food preservation as well as in medical fields. The downside of chitosan

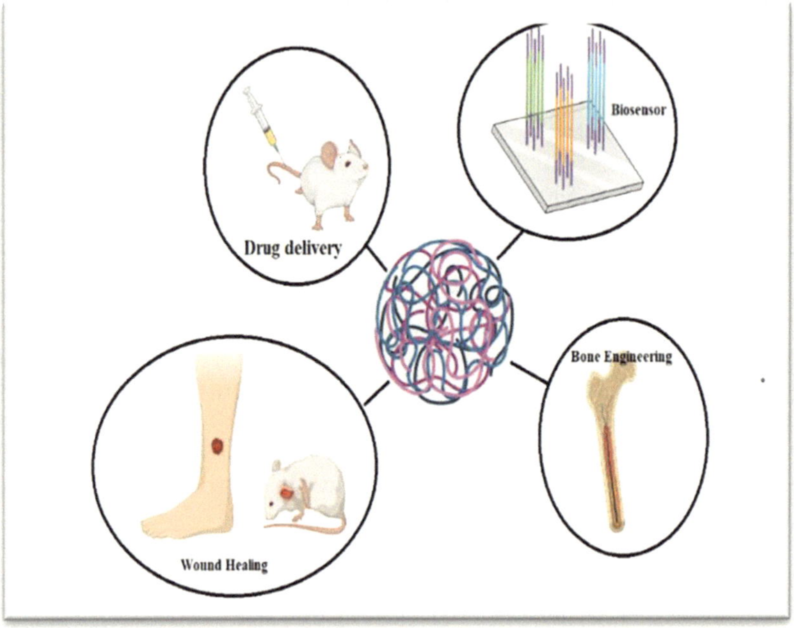

Figure 3.
Schematic representation of Chitosan-based nanocomposites in biomedical applications (https://biorender.com/).

is its solubility being in only an acidic solution. This is solved by using chitosan as a derivative of lactose to form chitilac [74, 75]. Burst release is less effective than the controlled release of drugs through nano carrier matrix-like nanofibers, membranes, spheres, capsules, tubes, etc.

The antimicrobial activity of chitosan nanocomposite [76] was utilized by Youssef. The effect of Montmorillonite-chitosan-silver sulfadiazine nanocomposites on skin lesions was also prepared and utilized by Sandri. Montmorillonite has hemostatic and absorbent features and it is used in the healing of lesions and ulcers [77, 78].

Wound healing is not the simple programmed structure of cellular and molecular developments including swelling, cell immigration, angiogenesis, temporal matrix syntheses, collagen deposition, and re-epithelization. Mostly, an effective wound dressing has the subsequent properties: (1) an appropriate water vapor transmission rate (WVTR), which produces a humid environment on the wound beds, without risking dehydration or exudates accumulation; (2) adequate gas penetrability for oxygen-requiring reparative methods; (3) a great level of fluid absorption ability to get rid of too many exudates, which cover nutrients for bacteria from the wound beds; (4) a good blockade trans the distribution of contagion producing microorganisms; (5) activity of antibacterial to overwhelm bacteria growth lower the dressing; and (6) the lack of any cytotoxic effects in the event of side damage to the neonatal tissues. Thus far numerous wound dressing materials types have been stated; however, they have some severe faults such as low water vapor/gas transmission rate, the capability of poor fluid absorption, and low flexible strength. So, we choose chitosan as a dressing material because of its biocompatibility [79–81] biodegradability [82], hemostatic activity [83, 84], the activity of anti-inflectional [85, 86], and property to accelerate wound healing [87–89]. The N-acetyl glucosamine (NAG) current in chitin and chitosan is a chief element of dermal tissue that is vital for the repair of scar tissue

[90, 91]. Its positive surface charge permits it to effectively support cell growth [92] and helps surface stimulated blood clotting and blood coagulation [93].

4.2 Tissue engineering applications of chitosan-based nanocomposites

3-D porous scaffolds for bone tissue studies should be biocompatible and let osteogenesis. Various techniques have been enlarged to imitate natural mineralized materials' properties and microstructure. Yet, large-size bulk materials' simple and fast fabrication with the content of high calcium under environmental aspects remains a big difficulty. Throughout gelation, a manageable inorganic gradient supply formula, accompanied by mineralization, where urea was used, and naturally hierarchically well-ordered hydrogel microstructures were shaped. This construction route takes a couple of hours to complete the gelation and processes of mineralization [94].

Composite scaffolds have been fabricated which use a coaxial electrospinning method to prepare gelatin-chitosan core-shell structured nanofibers. An arginine-glycine-aspartic acid (RGD)-like structure was shaped to imitate the structural component of the extracellular matrix of natural bone. After, by a wet chemical technique, hydroxyapatite was deposited on the prepared-shell structured nanofibers surface. Hydroxyapatite is the key mineral component of natural bone. Gelatin-chitosan core-shell structured nanofibers enhanced the mineralization productivity of hydroxyapatite compared to chitosan nanofibers [95]. Biologically-active scaffolds design is focused on cell-adhesive protein applications or bioceramic nanoparticles to produce a cell-sensitive surface. Hence, trace metals found in the alive organisms were used to prepare hydrogels of biocompatible chitosan. These were modified by copper (II) ions through complexation connections and generated a fewer constant cytocompatible to more constant cytotoxic structure for a copper-chitosan system [96, 97].

4.3 Cancer therapy applications of chitosan-based nanocomposites

Carcinoma involves the cell's uncontrolled proliferation. Operation and chemotherapy frequently are combined as a full strategy to defeat the tumor [98] Chemotherapy is a systemic treatment with the benefit of reducing lasting potential metastatic lesions after the operation. It has the benefit to treat numerous wounds contemporaneously, on the other hand, it is also apparent because of the systemic side effects, which may affect healthy tissue [99]. Numerous optimized chemotherapy strategies have been developed to solve the above-referred problems. Nanomaterials have been designed to target particular tissues or react to specific environmental conditions. CS is mostly developed to antitumor nanovesicles for the carcinoma behavior due to its matchless properties such as mucoadhesiveness and structural changefulness [100].

There are various articles about chitosan-based systems for antitumor drug delivery in advanced methods [101, 102]. Effective drug delivery systems are enhanced for therapy of anti-cancer based on environmental responsibility and directing principles to convey drugs, vaccines, etc. Furthermore, the drug delivery vectors were also designed with a mixture of photodynamic and hyperthermia cure (**Figure 4**) [103, 104].

Numerous healing anticancer drugs are limited in their scientific applications due to their toxicities and are not high solubility in aqueous media [105–107]. For example, doxorubicin (DOX) is one of the greatest commonly used drugs in cancer therapy. Yet, it can cause side reactions such as cardiotoxicity and drug resistance. Also, it is hard to manage intravenously due to its not high solubility in aqueous media.

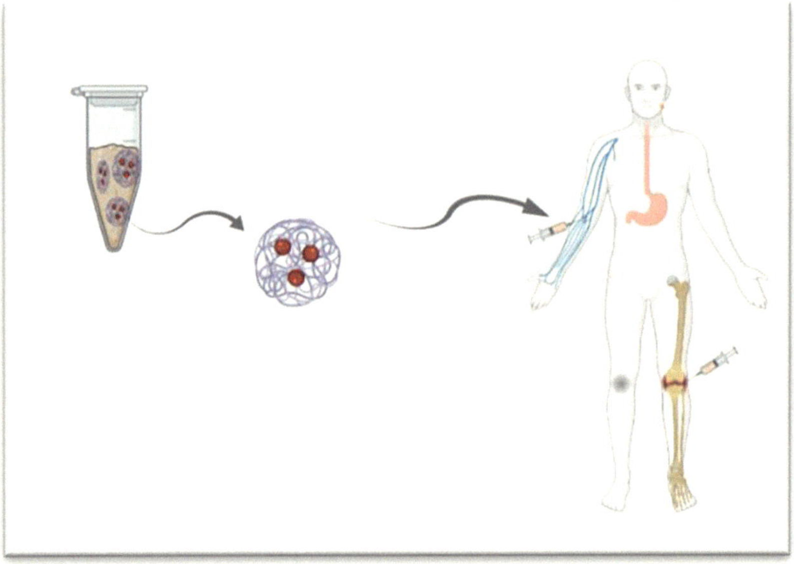

Figure 4.
Drug delivery (https://biorender.com/).

Nanomaterial-based drug delivery systems have received attention in overwhelming this downside. These systems can be made from several organic and inorganic materials including nondegradable and biodegradable polymers and inorganic nanocrystals. Polymeric micelles based on amphiphilic block copolymers have the benefits of great biocompatibility and capacity of drug-loading with less toxicity since they can self-assemble into polymeric micelles in aqueous media [108–110]. The mass in tumors through an improved penetration and retention (EPR) effect compared to single minor molecules, principal to special spatial distribution in the tumor. Nevertheless, the drug release performance of polymeric micelles is hard to control; they spontaneously release the drug before reaching tumors, which could give increased undesirable side effects and less therapeutic efficacy [111]. Well-designed drug delivery systems need to be advanced to enable cancer chemotherapy, which basically improves therapeutic efficacy by reducing drug release in unwanted sites. With these systems, a particular drug concentration can be delivered to tumors to diminish side effects. Drug delivery systems can be designed to release drugs stimulated by environmental parameters such as pH, enzymes, and temperature [112–114].

4.4 Gene delivery applications of chitosan-based nanocomposites

Gene therapy holds promise for challenging disease therapy such as innate genetic diseases [115], infections [116], and cancer [117]. A serious step for gene therapy is the gene's successful delivery. Naked nucleic acids cannot cross the cell membrane and are simply reduced by serum nucleases. Therefore, safe and efficient gene delivery systems are crucial for the accomplished application of gene therapy. Safety concerns about immunogenicity and toxicity are the key difficulties for the common use of viral systems [118]. Amongst virus-related vectors, adeno-associated virus (AAV) is now the most common gene delivery vector since it seldom includes the host genome, however, aggregate can be sensed [119, 120]. No viral vectors are good options for gene delivery because of their no genotoxicity, scalable manufacture, and structural elasticity [121]. They have been used in studies related to RNAi [122], gene editing [123], and CAR-T cell treatment [124]. No viral vectors are used more regularly in cancer treatment than in other uses for cancer and are a key worldwide public health matter [125]. Chitosan and its derivatives are commonly used to formulate gene delivery vectors. Various studies have been conducted to theory gene vectors based on chitosan. This assessment will argue the present development of policies encouraging the restrictions and chitosan applications in gene delivery systems and upcoming forecasts [126].

Gene therapy has developed very promptly due to its incredible healing possible to treat various genetic diseases by the inclusion of new genes (DNA and RNA) into target cells with transgene expression. Yet, the in vivo delivery of naked healing genes is unrealistic and tense with difficulties, including the gene susceptibility degradation by nucleases in the plasma, non-specificity to directed cells, in addition to the incapability of the negative genes from incoming negative cellular membranes. Certainly, with nearly 2600 gene therapeutics having finalized or ongoing medical trials, about six genes healing has received confirmation in the west [127, 128]. It is approved that the administration of the naked gene either systemically or locally is not effective, because of the concerns mentioned above. Consequently, the simple difficulties for gene treatment is the safe development and efficient gene transfection vector (**Figure 5**) [129].

4.5 Bio-imaging applications

Chitosan is a naturally happening amino-polysaccharide, with tempting physiochemical and biological features, gotten from the deacetylation of chitin, the second-highest biopolymer contemporary in the world afterward cellulose. The amino and hydroxyl groups current on the skeleton of chitosan deliver a route for reaction with

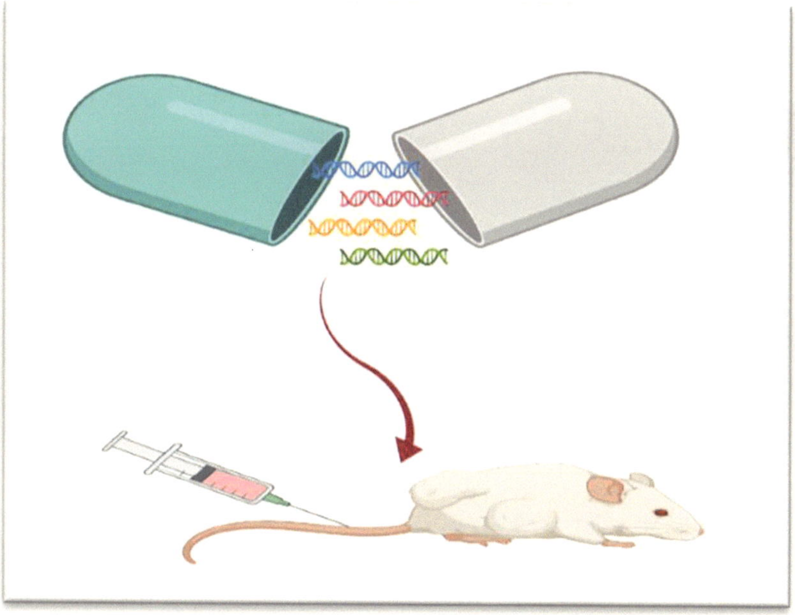

Figure 5.
Gene delivery in the cancer therapy (https://biorender.com/).

organic practical molecules [130]. Moreover, chitosan is used as a biomaterial due to its outstanding biodegradability and low toxicity. Also used as bio imaging agent [131]. The introduction of imaging agents in chitosan permitted its use for bioimaging. Such as the integration of imaging agents, for example, Fe_3O_4 figure magnetic resonance imaging; the self-assembled nanoparticles improve the perception of hepatocyte targeted imaging [132]. Chitosan decreased gold nanoparticles have performed as photo thermal-converter, photodynamic-treatment besides photodynamics carrier and thus figure in the bio-imaging implementation plus used to annihilate the breast cancer cells (MCF-7) [133]. Gold-covered Fe_3O_4 nanoparticles were made by chemical co-precipitation technique with an average size of 9.8 nm in diameter by the decrease of glucose and stabilized with chitosan in the face of formaldehyde as a crosslinking agent. The prepared nanomaterial performed as well substantial for bio establisher applications [134, 135].

4.6 Applications of biosensor

Biosensors are equipment to transform biological, physical, and chemical indications of biological schemes into electrical by classifying particular responses to aim analyses [136]. A blood glucose biosensor is a typical sample of a characteristic biosensor that uses the enzyme Glucose oxidase. Electrochemical biosensors particularly react with target moieties and produce an electrical signal linked to particular analyze concentrations, pH, and temperature [137]. Biosensors are a vital role in tissue engineering, and also chitosan nanocomposites are well worn in these kinds of applications. Generally, metal nanoparticles display higher conductivity and electronic features than transmission polymer, however, the flexibility of the polymer creates them matchless for numerous applications. The addition of transmission

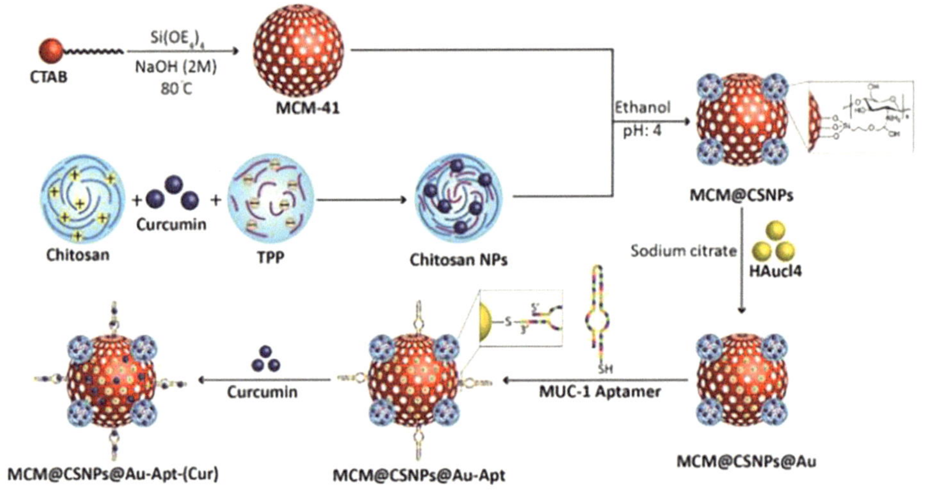

Figure 6.
Stepwise schematic of the nanosystem synthesis: MCM-41 was prepared through a surfactant-assisted sol-gel method, then, the fabricated chitosan-curcumin nanoparticles (CSNPs) were crosslinked onto MCM-41 through TPP (MCM@CS). In the next step, the synthesized AuNPs were modified physically onto the MCM-41 surface as well as loaded into CSNPs networks (MCM@CS@Au), and MUC-1 aptamer as a targeted agent linked to AuNPs via sulfuric bonds (MCM@CS@Au-Apt); eventually curcumin was loaded into the final nanosystem (MCM@CS@Au-Apt (CUR)) [147] (Esmaeili Y, Khavani M, Bigham A, Sanati A, Bidram E, Shariati L, Zarrabi A, Jolfaie NA, Rafienia M. Mesoporous silica@chitosan@gold nanoparticles as "on/off" optical biosensor and pH-sensitive theranostic platform against cancer. Int J Biol Macromol. 2022 Mar 31;202:241–255).

Materials	References
Laccase-Based Biosensor Encapsulated in a Galactomannan-Chitosan Composite	[148]
Ti 3 C 2 T X MXene/Chitosan Nanocomposite-Based Amperometric Biosensor	[149]
Glucose oxidase-chitosan immobilized paper biosensor	[150]
Diffraction-based chitosan leaky waveguide (LW) biosensors	[151]
Carboxymethyl chitosan assembled piezoelectric biosensor	[152]
Au nanoparticles/MoS 2 nanosheets—Chitosan modified screen-printed electrode as chlorpyrifos biosensor	[153]
Temperature-sensitive poly(N-isopropylacrylamide)-chitosan hydrogel for fluorescence sensors	[154]

Table 1.
Chitosan-based biosensor applications.

nanoparticles to chitosan materials increases the nanocomposite electrical conductivity on top of stimuli-responsive features that could be used for sensing biological types [138]. For particularly improving the sensitivity to biological moieties, biosensor faces have been changed using enzymes similar to cholesterol esterase [139] and cholesterol oxidase [140] to determine cholesterol substance in the blood or human serums. As it is seen chitosan nanocomposites-based biosensors are higher efficient, sensitive, and durable in comparison to pure chitosan [138]. A lot of nanostructured inorganic materials for instance cuprous oxide nanoparticles [141] and Fe_3O_4 nanoparticles [142]. $NiFe_2O_4$ nanoparticle [143], Cerium oxide nanoparticle [144], and TiO_2 nanoparticles [145] are often used to increase the electronic features, moreover, the chitosan electrical conductivity-based materials in nanocomposites [146].

In another study, Esmaelili *et al.* developed a multifunctional nanosystem of mesoporous silica@chitosan@gold-aptamer (MCM@CS@Au-Apt) encapsulated curcumin for MUC-1 positive tumor cells targeting and drug delivery (**Figure 6**). They found that MCM@CS@Au has "on/off" fluorescence bio-sensing ability with selective performance against MUC-1 positive tumor cells

Table 1 showed chitosan-based biosensor applications in the recent years

5. Conclusion and future perspectives

Chitosan is a natural polymer and has many advantages including nontoxicity, biocompatibility, and biodegradability, and it can be applied in many fields, especially in medicine. The importance of chitosan-based bionanocomposites is skyrocketing due to their numerous advantages. Because the use and potential of chitosan-based bionanocomposite expanding in a variety of industries, this paper provides a thorough examination of these issues. The negative effects of non-biodegradable materials on humans and the environment have sparked interest in biodegradable materials among scientists and environmentalists. As a result of its intrinsic qualities such as biocompatibility, biodegradability, and nontoxicity, as well as their improved structural and functional aspects, bionanocomposites have become a focus of considerable research. Bionanocomposites based on chitosan are the most popular at the moment. They have a lot of great features and advantages. This chapter primarily focuses on the properties of chitosan, and applications of chitosan-based nanocomposites. In summary, nanocomposites based on chitosan have great potential for research and development of new nanomaterials in the future.

Scope of the review and its significance

Recent research on chitosan-based nanocomposites for biomedical applications is based on the field's expanding understanding of chitosan properties and chemical/physical alteration. This chapter discussed the current understanding of the effect of chitosan-based nanocomposites in medical studies.

Translational relevance

Usage of polymer-based nanomaterials has been analyzed for their varied application and the most important of these is the use of chitosan-based polymeric nanomaterial in the medical field. Studies confirm that the use of chitosan-based materials in the clinic is increasing day by day (https://clinicaltrials.gov).

Clinical relevance

In the chapter, the chitosan-based nanocomposites and their biomedical applications in the biomedical field involving drug delivery systems and other applications have been addressed highlighting the recent advancements.

Author disclosure and ghostwriting

There are no competing interests to declare. No ghostwriters were used to write this article.

Abbreviations

DD	Deacetylation degree
Mw	Molecular Weight
MMT	Montmorillonite
organoMMT	Cloisite30B/15A
FTIR	Fourier-transform infrared spectroscopy
XRD	X-ray Diffraction Analysis
SEM	Scanning Electron Microscopes
DSC	Differential scanning calorimetry
WVTR	Water vapor transmission rate
NAG	N-acetyl glucosamine
DOX	Doxorubicin
EPR	Penetration and retention effect
AAV	Adeno-associated virus

DOI: http://dx.doi.org/10.5772/intechopen.106379

Author details

Serap Yalcin[1*] and Nevin Cankaya[2]

1 Department of Molecular Biology and Genetic, Kırşehir Ahi Evran University, Kırşehir, Turkey

2 Department of Chemistry, Uşak University, Uşak, Turkey

*Address all correspondence to: syalcin@ahievran.edu.tr

References

[1] Çankaya N, Şahin R. Chitosan/clay bionanocomposites: Structural, antibacterial, thermal and swelling properties. Cellulose Chemistry and Technology. 2019;**53**(5-6):537-549

[2] Domínguez-Delgado CL, Rodríguez-Cruz IM, Fuentes-Prado E, Escobar-Chávez JJ, Vidal-Romero G, García-González L, et al. . In: Gowder SJT, editor. Pharmacology and Therapeutics: Drug Carrier Systems Using Chitosan for Non Parenteral Routes. 5th ed. Rijeka, Croatia: InTechOpen Limited; 2014. pp. 273-325

[3] Safdar R, Omar AA, Arunagiri A, Regupathi I, Thanabalan M. Potential of chitosan and its derivatives for controlled drug release applications–A review. Journal of Drug Delivery Science and Technology. 2019;**49**:642-659

[4] İnal M. Controlled release of indomethacin from chitosan/κ-carrageenan/chitosan three layer microspheres. Fırat University Journal of Engineering Science. 2020;**32**(2):425-433

[5] Çankaya N. Grafting of chitosan: Structural, thermal and antimicrobial properties. Chemical Society of Pakistan. 2019;**41**(2):240-245

[6] Berger J, Reist M, Mayer JM, Felt O, Peppas NA, Gurny R. Structure and interactions in covalently and ionically crosslinked chitosan hydrogels for biomedical applications. European Journal of Pharmaceutics and Biopharmaceutics. 2004;**57**:19-34

[7] Çankaya N, Sökmen Ö. Chitosan-clay bionanocomposites. Journal of Polytechnic. 2016;**19**(3):283-295

[8] Cho J, Heuzey MC, Bégin A, Carreau PJ. Physical gelation of chitosan in the presence of α- glycerophosphate: The effect of temperature. Biomacromolecules. 2005;**6**:3267-3275

[9] Demir A, Seventekin N. Chitin, chitosan and general application areas. Electronic Journal of Text and Technology. 2009;**3**(2):92-103

[10] Rao MS, Stevens WF. Chitin production by lactobacillus fermentation of shrimp biowaste in a drum reactor and its chemical conversion to chitosan. Journal of Chemical Technology and Biotechnology. 2005;**80**:1080-1087

[11] Kurita K. Chitin and chitosan: Functional biopolymers from marine crustaceans. Biotechnology. 2006;**8**:203-226

[12] Anthonsen MW, Varum KM, Smidsrod O. Solution properties of chitosans: Conformation and chain stiffness of chitosans with different degrees of N-acetylation. Carbohydrate Polymer. 1993;**2**:193-201

[13] Hejazi R, Amiji M. Chitosan-based gastrointestinal delivery systems. Journal of Controlled Release. 2003;**89**:151-155

[14] Philippova OE, Korchagina EV, Volkov EV, Smirnov VA, Khokhlov AR, Rinaudo M. Aggregation of some water-soluble derivatives of chitin in aqueous solutions: Role of the degree of acetylation and effect of hydrogen bond breaker. Carbohydrate Polymers. 2012;**87**:687-694

[15] Baxter A, Dillon M, Taylor KDA, Roberts GAF. Improved method for i.r. determination of the degree of N-acetylation of chitosan. International Journal of Biological Macromolecules. 1992;**14**(3):166-169

[16] Wenling C, Duohui J, Jiamou L, Yandao G, Nanming Z, Xiufang Z. Effects of the degree of deacetylation on the physicochemical properties and Schwann cell affinity of chitosan films. Journal of Biomaterials Applications. 2005;**20**(2):157-177

[17] Khor E, Lim LY. Implantable applications of chitin and chitosan. Biomaterials. 2003;**24**(13):2339-2349

[18] Zambaux MF, Bonneaux F, Gref R, Dellacherie E, Vigneron C. Preparation and characterization of protein C-loaded PLA nanoparticles. Journal of Controlled Release. 1999;**60**(2-3):179-188

[19] Kimna C, Deger S, Tamburaci S, Tihminlioglu F. Chitosan/montmorillonite composite nanospheres for sustained antibiotic delivery at post-implantation bone infection treatment. Biomedical Materials. 2019;**14**:044101

[20] No HK, Lee SH, Park NY, Meyers SP. Comparison of physicochemical, binding, and antibacterial properties of chitosans prepared without and with deproteinization process. Journal of Agricultural and Food Chemistry. 2003;**51**:7659-7663

[21] Rinaudo M. Chitin and chitosan: Properties and applications. Progress in Polymer Science. 2006;**31**:603-632

[22] Croisier F, Jérôme C. Chitosan-based biomaterials for tissue engineering. European Journal of Polymer Science. 2013;**49**:780-792

[23] Çankaya N. Grafting studies of chitin. Sigma Journal of Engineering and Natural Science. 2019;**37**(1):111-117

[24] Çankaya N. Synthesis and structural properties of chitin/clay bio-nanocomposites. Cellulose Chemistry and Technology. 2021;**55**(5-6):659-665

[25] Gürsoy AZ, editor. Kontrollü Salım Sistemleri. 2nd ed. İstanbul, Turkey: Kontrollü Salım Sistemleri Derneği Yayını; 2014

[26] Gürsoy AZ, editor. Farmasötik Teknoloji. 2nd ed. İstanbul, Turkey: Kontrollü Salım Sistemleri Derneği Yayını; 2012

[27] Tüylek Z. Drug delivery systems and nanotechnological interaction. Bozok Medical Journal. 2017;**7**(3):89-98

[28] Fialho SL, Behar-Cohen F, Silva-Cunha A. Dexamethasone-loaded poly(ε-caprolactone)intravitreal implants: A pilot study. European Journal of Pharmaceutics and Biopharmaceutics. 2007;**68**:637-646

[29] Lee ES, Schwartz FW. Characteristics and applications of controlled–release KMnO 4 for groundwater remediation. Chemosphere. 2007;**66**:2058-2066

[30] Han X, Chen S, Hu X. Controlled-release fertilizer encapsulated by starch/polyvinyl alcohol coating. Desalination. 2009;**240**:21-26

[31] Ahmed I, Kasraian K. Pharmaceutical challenges in veterinary product development. Advanced Drug Delivery Reviews. 2002;**54**:871-882

[32] Kost J, Langer R. Responsive polymeric delivery systems. Advanced Drug Delivery Reviews. 2001;**46**:125-148

[33] Qiu Y, Park K. Environment-sensitive hydrogels for drug delivery. Advanced Drug Delivery Reviews. 2001;**53**:321-339

[34] Read NW, Sugden K. Gastrointestinal dynamics and pharmacology for the optimum desing of controlled release oral dosage forms. Critical Reviews in Therapeutic Drug Carrier Systems. 1987;**4**:221-263

[35] Gürsoy AZ, editor. Nanofarmasötikler ve Uygulamaları. 1st ed. İstanbul, Turkey: Kontrollü Salım Sistemleri Derneği Yayını; 2014

[36] Bajpai AK, Shukla SK, Bhanu S, Kankane S. Responsive polymers in controlled drug delivery. Progress in Polymer Science. 2008;**33**:1088-1118

[37] Brannon-Peppas L. Polymers in controlled drug delivery. Medical Plastic and Biomaterials. 1997;**4**:34-45

[38] Erşan Y, Önder S. Synthesis of chitosan based drug delivery system and evaluation its performance in vitro. European Journal of Science and Technology. 2020;**20**:270-279

[39] Cam BN, Genisel M. Overview of the targeted treatment through nano-drug delivery systems and controlled-drug release systems. Eastern Anatolian Journal of Science. 2019;**5**(1):43-49

[40] Tandya A, Mammucari R, Dehghani F, Foster NR. Dense gas processing of polymeric controlled release formulations. International Journal of Pharmaceutics. 2007;**328**(1):1-11

[41] Kim C, editor. Controlled Release Dosage Form Design. 1st ed. Florida: CRC Press LLC; 2000

[42] Tüylek Z. Drug delivery nanosystems. Archives Medical Review Journal. 2019;**28**(2):184-192

[43] Yuan Q, Shah J, Hein S, RDK M. Controlled and extended drug release behavior of chitosan- based nanopasticle carrier. Acta Biomaterialia. 2010;**6**:1140-1148

[44] Hua S, Yang H, Wang W, Wang A. Controlled release of oflaxacin from chitosan-montmorillonite hydrogel. Applied Clay Science. 2010;**50**:112-117

[45] Cheikh D, Garcia-Villen F, Majdoub H, Viseras C, Zayani MB. Chitosan/beidellite nanocomposite as diclofenac carrier. International Journal of Biological Macromolecules. 2019;**129**:44-53

[46] Sharma A, Puri V, Kakkar V, Singh I. Formulation and evaluation of silymarin-loaded chitosan- montmorillonite microbeads for the potential treatment of gastric ulcers. Journal of Functional Biomaterials. 2018;**9**(3):52

[47] Depan D, Kumar AP, Singh RP. Cell proliferation and controlled drug release studies ofnanohybrids based on chitosan-g-lactic acid and montmorillonite. Acta Biomaterialia. 2009;**5**:93-100

[48] Sahoo S, Sasmal A, Sahoo D, Nayak P. Synthesis and characterization of chitosan-polycaprolactone blended with organoclay for control release of doxycycline. Journal of Applied Polymer Science. 2010;**118**:3167-3175

[49] Parida UK, Nayak AK, Binhani BK, Nayak PL. Synthesis and characterization of chitosan- polyvinyl alcohol blended with Cloisite 30B for controlled release of the anticancer drug curcumin. Journal of Biomaterials and Nanobiotechnology. 2011;**2**:414-425

[50] Nanda R, Sasmal A, Nayak PL. Preparation and characterization of chitosan–polylactidecomposites blended with Cloisite 30B for control release of the anticancer drug paclitaxel. Carbohydrate Polymers. 2011;**83**:988-994

[51] Cojocariu A, Profire L, Aflori M, Vasile C. In vitro drug release from chitosan/Cloisite 15A hydrogels. Applied Clay Science. 2012;**57**:1-9

[52] Çaykara T, Akçakaya I. Synthesis and network structure of ionic poly(N,N-dimethylacrylamide- co-acrylamide) hydrogels: Comparison of swelling

degree with theory. European Polymer Journal. 2006;**42**(6):1437-1445

[53] Wang Q, Zhang JZ, Wang A. Preparation and characterization of a novel pH sensitive chitosan-g-poly(acrylic acid)/attapulgite/sodium alginate composite hydrogel bead for controlled release of diclofenac sodium. Carbohydrate Polymers. 2009;**78**:731-737

[54] Dinu MV, Cocarta AI, Dragan ES. Synthesis, characterization and drug release properties of 3D chitosan/clinoptilolite biocomposite cryogels. Carbohydrate Polymers. 2016;**153**:203-211

[55] Luo Y, Mills DK. The effect of halloysite addition on the material properties of chitosan–halloysite hydrogel composites. Gels. 2019;**5**:40

[56] Ma L, Liu M, Liu H, Chen J, Gao C, Cui D. Dual crosslinked pH- and temperature-sensitive hydrogel beads for intestine-targeted controlled release. Polymers for Advanced Technologies. 2010;**21**(5):348-355

[57] Aycan D, Alemdar N. Development of pH-responsive chitosan-based hydrogel modified with bone ash for controlled release of amoxicillin. Carbohydrate Polymers. 2018;**184**:401-407

[58] Yücel Ç, Karatoprak GŞ, Aktaş Y, Koşar M. In vitro evaluatıon of quercetin-loaded chitosan nanoparticles. Journal of Health Science Erciyes University. 2016;**25**:41-44

[59] Ulu A. Preparation of allantoin loaded chitosan nanoparticles and influence of molecular weight of chitosan on drug release. Journal of Inonu University and Health Services Vocational School. 2020;**8**(3):725-740

[60] Önder S. Hastane enfeksiyonlarına karşı implant yüzeylerinden kontrollü ilaç salami. Gazi University Journal of Science. 2016;**4**(3):127-134

[61] Kumar CSSR, Mohammad F. Magnetic nanomaterials for hyperthermia-based therapy and controlled drug delivery. Advanced Drug Delivery Reviews. 2011;**63**:789-808

[62] Zhang Y, Kohler N. Zhang M, Surface modification of superparamagnetic magnetite nanoparticles and their intracellular uptake. Biomaterials. 2002;**23**:1553-1561

[63] Zrínyi M, Barsi L, Büki A. Ferrogel: A new magneto-controlled elastic medium. Polymer Gels Networks. 1997;**5**:415-427

[64] Mahdavinia GR, Mosallanezhad A, Solaymani M, Sabzica M. Magnetic- and pH-responsive κ- carrageenan/chitosan complexes for controlled release of methotrexate anticancer drug. International Journal of Biological Macromolecules. 2017;**97**:209-217

[65] Long J, Yu X, Xu E, Wu Z, Xu X, Jin Z, et al. In situ synthesis of new magnetite chitosan/carrageenan nanocomposites by electrostatic interactions for protein delivery applications. Carbohydrate Polymers. 2015;**131**:98-107

[66] Wang D, Li J, Li H, Tang F. Preparation and drug releasing property of magnetic chitosan-5- fluorouracil nano-particles. Transactions of Nonferrous Metals Society of China. 2009;**19**(5):1232-1236

[67] Sivashankari P, Prabaharan M. Prospects of chitosan-based sca_olds for growth factor release in tissue engineering. International Journal of Biological Macromolecules. 2016;**93**:1382-1389

[68] Misgav M, Lubetszki A, Brutman-Barazani T, Martinowitz U, Kenet G. The hemostatic efficacy of

chitosan-pads in hemodialysis patients with significant bleeding tendency. The Journal of Vascular Access. 2017;**18**: 220-224

[69] Ahmed TA, Aljaeid BM. Preparation, characterization, and potential application of chitosan, chitosan derivatives, and chitosan metal nanoparticles in pharmaceutical drug delivery. Drug Design, Development and Therapy. 2016;**10**:483

[70] Yang H, Chen Y, Chen Z, Geng Y, Xie X, Shen X, et al. Chemo-photodynamiccombined gene therapy and dual-modal cancer imaging achieved by pH-responsive alginate/chitosan multilayer-modifiedmagneticmesoporous silica nanocomposites. Biomaterials Science. 2017;**5**:1001-1013

[71] Sengiz C, Congur G, Eksin E, Erdem A. Multiwalled carbon nanotubes-chitosan modified single-use biosensors for electrochemical monitoring of drug-DNA interactions. Electroanalysis. 2015;**27**:1855-1863

[72] Lee H, Kim M, Yoon Y, Park W. Fluorescent property of chitosan oligomer and its application as a metal ion sensor. Marine Drugs. 2017;**15**:105

[73] Rodrigues SV, John LE, Mitra DK, Shah R, Shetty G, Prithyani S, et al. Evaluation and comparison of antimicrobial effects of chlorhexidine (CHX) and chitosan (CHT) mouthwash in chronic periodontitis (CGP) patients-A Clinico-microbiological Study. JIDA Journal of Indian Dental Association. 2018;**16**:26-32

[74] Bano I, Arshad M, Yasin T, Ghauri MA, Younus M. Chitosan: A potential biopolymer for wound management. International Journal of Biological Macromolecules. 2017;**102**: 380-383

[75] Batrakova EV, Gendelman HE, Kabanov AV. Cell-mediated drug delivery. Expert Opinion in Drug Delivery. 2011;**8**(4):415-433

[76] Youssef AM, Abou-Yousef H, El-Sayed SM, Kamel S. Mechanical and antibacterial properties of novel high performance chitosan/nanocomposite films. International Journal of Biological Macromolecules. 2015;**76**:25-32

[77] Salcedo I, Aguzzi C, Sandri G, Bonferoni MC, Mori M, Cerezo P, et al. In vitro biocompatibility and mucoadhesion of montmorillonite chitosan nanocomposite: A new drug delivery. Applied Clay Science. 2012;**55**:131-137

[78] Sandri G, Bonferoni MC, Ferrari F, Rossi S, Aguzzi C, Mori M, et al. Carbohydrate Polymers. n.d.;**102**:970

[79] Archana D, Dutta J, Dutta PK. Evaluation of chitosan nano dressing for wound healing: Characterization, in vitro and in vivo studies. International Journal of Biological Macromolecules. 2013;**57**:193-203

[80] Madhumathi K, Shalumon KT, Rani VV, et al. Wet chemical synthesis of chitosan hydrogel-hydroxyapatite composite membranes for tissue engineering applications. International Journal of Biological Macromolecules. 2009;**45**(1):12-15

[81] Jayakumar R, Menon D, Manzoor K, Nair SV, Tamura H. Biomedical applications of chitin and chitosan based nanomaterials—A short review. Carbohydrate Polymers. 2010;**82**(2):227-232

[82] Singh J, Dutta PK. Preparation, circular dichroism induced helical conformation and optical property of chitosan acid salt complexes for biomedical applications. International

Journal of Biological Macromolecules. 2009;**45**(4):384-392

[83] de Queiroz AA, Ferraz HG, Abraham GA, del Mar Fernández M, Bravo AL, Román JS. Development of new hydroactive dressings based on chitosan membranes: Characterization and in vivo behavior. Journal of Biomedical Materials Research Part A. 2003;**64**(1):147-154

[84] Li X, Kong X, Zhang Z, et al. Cytotoxicity and biocompatibility evaluation of N,O-carboxymethyl chitosan/oxidized alginate hydrogel for drug delivery application. International Journal of Biological Macromolecules. 2012;**50**(5):1299-1305

[85] Archana D, Dutta J, Dutta PK. Dutta Chitosan–pectin–titanium dioxide nanocomposite film: An investigation for wound healing applications. Asian Chitin Journal. 2013;**57**:193-203

[86] Luca L, Rougemont AL, Walpoth BH, et al. Injectable rhBMP-2-loaded chitosan hydrogel composite: Osteoinduction at ectopic site and in segmental long bone defect. Journal of Biomedical Materials Research Part A. 2011;**96**(1):66-74

[87] Alsarra IA. Chitosan topical gel formulation in the management of burn wounds. International Journal of Biological Macromolecules. 2009;**45**(1):16-21

[88] Xu H, Ma L, Shi H, Gao C, Han C. Chitosan–hyaluronic acid hybrid film as a novel wound dressing: In vitro and in vivo studies. Polymers for Advanced Technologies. 2007;**18**(11):869-875

[89] Kumar PT, Lakshmanan VK, Anilkumar TV, et al. Flexible and microporous chitosan hydrogel/nano ZnO composite bandages for wound dressing: In vitro and in vivo evaluation [retracted in: ACS Appl Mater Interfaces. 2019 Aug 7;**11**(31):28596]. ACS Applied Materials & Interfaces. 2012;**4**(5):2618-2629

[90] Singh DK, Ray AR. Biomedical applications of chitin, chitosan, and their derivatives. Journal of Macromolecular Science, Part C: Polymer Reviews. 2000;**40**(1):69-83

[91] Zielinski BA, Aebischer P. Chitosan as a matrix for mammalian cell encapsulation. Biomaterials. 1994;**15**(13):1049-1056

[92] Amiji MM. Permeability and blood compatibility properties of chitosan-poly(ethylene oxide) blend membranes for haemodialysis. Biomaterials. 1995;**16**(8):593-599

[93] Cheng B, Pei B, Wang Z, Hu Q. Advances in chitosan-based superabsorbent hydrogels. RSC Advances. 2017;**7**:42036-42046

[94] Kravanja G, Primožič M, Knez Ž, Leitgeb M. Chitosan-based (Nano)materials for Novel Biomedical Applications. Molecules. 2019;**24**(10):1960

[95] Cui Z, Zheng Z, Lin L, Si J, Wang Q, Peng X, et al. Electrospinning and crosslinking of polyvinyl alcohol/chitosan composite nanofiber for transdermal drug delivery. Advances in Polymer Technology. 2018;**37**:1917-1928

[96] Rogina A. Preparation and characterization of nano-hydroxyapatite within chitosan matrix. Materials Science and Engineering: C. 2013;**33**:4539-4544

[97] Ressler A, Ródenas-Rochina J. Injectable chitosanhydroxyapatite hydrogels promote the osteogenic di_ erentiation of mesenchymal stem cells. Carbohydrate Polymer. 2018;**197**:469-477

[98] Ye Z, Wu WR, Qin Y-F, Hu J, Liu C, Seeberger PH, et al. An integrated therapeutic delivery system for enhanced treatment of hepatocellular carcinoma. Advanced Functional Materials. 2018;**28**: 1706600

[99] Zhang X, Xu G, Gadora K, Cheng H, Peng J, Ma Y, et al. Dual-sensitive chitosan derivative micelles for site-specific drug release in the treatment of chicken coccidiosis. RSC Advances. 2018;**8**:14515-14526

[100] Li J, Cai C, Li J, Li J, Li J, Sun T, et al. Chitosan-based nanomaterials for drug delivery. Molecules. 2018;**23**(10):2661. DOI: 10.3390/molecules23102661

[101] Fathi M, Majidi S, Zangabad PS, Barar J, Erfan-Niya H, Omidi Y. Chitosan-based multifunctional nanomedicines and theranostics for targeted therapy of cancer. Medicinal Research Reviews. 2018;**38**:2110-2136

[102] Cosco D, Cilurzo F, Maiuolo J, Federico C, Di Martino MT, Cristiano MC, et al. Delivery of miR-34a by chitosan/PLGA nanoplexes for the anticancer treatment of multiple myeloma. Scientific Reports. 2015;**5**:17579-17589

[103] Assa F, Jafarizadeh-Malmiri H, Ajamein H, Vaghari H, Anarjan N, Ahmadi O, et al. Chitosan magnetic nanoparticles for drug delivery systems. Critical Reviews in Biotechnology. 2017;**37**:492-509

[104] Li F, Mei H, Gao Y, Xie X, Nie H, Li T, et al. Co-delivery of oxygen and erlotinib by aptamer-modified liposomal complexes to reverse hypoxia-induced drug resistance in lung cancer. Biomaterials. 2017;**145**:56-71

[105] Janib SM, Moses AS, MacKay JA. Imaging and drug delivery using theranostic nanoparticles. Advanced Drug Delivery Reviews. 2010;**62**:1052-1063

[106] Cho H, Dong Z, Pauletti G, Zhang J, Xu H, Gu H, et al. Fluorescent, superparamagnetic nanospheres for drug storage, targeting, and imaging: A multifunctional nanocarrier system for cancer diagnosis and treatment. ACS Nano. 2010;**4**:5398-5404

[107] Wang J, Sun X, Mao W, Sun W, Tang J, Sui M, et al. Tumor redox heterogeneity-responsive prodrug nanocapsules for cancer chemotherapy. Advanced Materials. 2013;**25**:3670-3676

[108] Lim E-K, Jang E, Lee K, Haam S, Huh Y-M. Delivery of cancer therapeutics using nanotechnology. Pharmaceutics. 2013;**5**:294-317

[109] Kamada H, Tsutsumi Y, Yoshioka Y, Yamamoto Y, Kodaira H, et al. Design of a pH-sensitive polymeric carrier for drug release and its application in cancer therapy. Clinical Cancer Research. 2004;**10**:2545-2550

[110] Zhang CY, Yang YQ, Huang TX, Zhao B, Guo XD, Wang JF, et al. Self-assembled pH-responsive MPEG-b-(PLA-co-PAE) block copolymer micelles for anticancer drug delivery. Biomaterials. 2012;**33**:6273-6283

[111] Secret E, Smith K, Dubljevic V, Moore E, Macardle P, Delalat B, et al. Antibody-functionalized porous silicon nanoparticles for vectorization of hydrophobic drugs. Advanced Healthcare Materials. 2013;**2**:718-727

[112] Prabaharan M, Grailer JJ, Pilla S, Steeber DA, Gong S. Amphiphilic multi-arm-block copolymer conjugated with doxorubicin via pH-sensitive hydrazone bond for tumor-targeted drug delivery. Biomaterials. 2009;**30**:5757-5766

[113] Kosif I, Cui M, Russell TP, Emrick T. Triggered in situ disruption and inversion of nanoparticle-stabilized droplets. Angewandte Chemie (International Ed. in English). 2013;**52**:6620-6623

[114] Lim EK, Sajomsang W, Choi Y, Jang E, Lee H, Kang B, et al.

Chitosan-based intelligent theragnosis nanocomposites enable pH-sensitive drug release with MR-guided imaging for cancer therapy. Nanoscale Research Letters. 2013;**8**(1):467

[115] Gao X, Tao Y, Lamas V, Huang M, Yeh WH, Pan B, et al. Treatment of autosomal dominant hearing loss by in vivo delivery of genome editing agents. Nature. 2018;**553**:217-221

[116] Hetzel M, Mucci A, Blank P, Nguyen AHH, Schiller J, Halle O, et al. Hematopoietic stem cell gene therapy for IFN gamma R1 deficiency protects mice from mycobacterial infections. Blood. 2018;**131**:533-545

[117] Zhao W, Yang Y, Song L, Kang T, Du T, Wu Y, et al. A vesicular stomatitis virus-inspired DNA nanocomplex for ovarian cancer therapy. Advanced Science. 2018;**5**(3):1700263

[118] Song H, Yu MH, Lu Y, Gu ZY, Yang YN, Zhang M, et al. Plasmid DNA delivery. Journal of the American Chemical Society. 2017;**139**:18247-18254

[119] Shi J, Zheng DX, Liu YX, Sham MH, Tam P, Farzaneh F, et al. Overexpression of soluble trail induces apoptosis in human lung adenocarcinoma and inhibits growth of tumor xenografts in nude mice. Cancer Research. 2005;**65**:1687-1692

[120] Yin H, Song CQ, Suresh S, Wu QQ, Walsh S, Rhym LH, et al. Structure-guided chemical modification of guide RNA enables potent non-viral in vivo genome editing. Nature Biotechnology. 2017;**35**:1179-1187

[121] Cortez MA, Godbey WT, Fang YL, Payne ME, Cafferty BJ, Kosakowska KA, et al. The synthesis of cyclic poly(ethylene imine) and exact linear analogues: An evaluation of gene delivery comparing polymer architectures. Journal of the American Chemical Society. 2015;**137**:6541-6549

[122] Rietwyk S, Peer D. Next-generation lipids in RNA interference therapeutics. ACS Nano. 2017;**11**:7572-7586

[123] Lee K, Conboy M, Park HM, Jiang FG, Kim HJ, Dewitt MA, et al. Nanoparticle delivery of cas9 ribonucleoprotein and donor DNA in vivo induces homology-directed DNA repair. Nature Biomedical Engineering. 2017;**1**:889-901

[124] Smith TT, Stephan SB, Moffett HF, McKnight LE, Ji WH, Reiman D, et al. In situ programming of leukaemia-specific T cells using synthetic DNA nanocarriers. Nature Nanotechnology. 2017;**12**:813-820

[125] Gao X, Wang S, Wang B, Deng S, Liu X, Zhang X, et al. Improving the anti-ovarian cancer activity of docetaxel with biodegradable self-assembly micelles through various evaluations. Biomaterials. 2015;**53**:646-658

[126] Chuan D, Jin T, Fan R, Zhou L, Guo G. Chitosan for gene delivery: Methods for improvement and applications. Advances in Colloid and Interface Science. 2019;**268**:25-38

[127] Anguela XM, High KA. Entering the modern era of gene therapy. Annual Review of Medicine. 2019;**70**: 273-288. DOI: 10.1146/annurev-med-012017-043332

[128] Ginn SL, Amaya AK, Alexander IE, Edelstein M, Abedi MR. Gene therapy clinical trials worldwide to 2017: An update. The Journal of Gene Medicine. 2018;**20**:e3015. DOI: 10.1002/jgm.3015

[129] Cao Y, Tan YF, Wong YS, Liew MWJ, Venkatraman S. Recent advances in chitosan-based carriers for gene delivery. Marine Drugs. 2019;**17**:381

[130] Martin GL, Ross J, Minteer SD, Jamesonb DM, Cooney MJ. Fluorescence characterization of chemical micro-environments in hydrophobically modified chitosan. Carbohydrate Polymers. 2009;77:695

[131] van de Manakker F, Vermonden T, van Nostrum CF, Hennink WE. Bio Cyclodextrin-based polymeric materials: Synthesis, properties, and pharmaceutical/biomedical applications. Biomacromolecules. 2009;**10**:3157

[132] Lee C-M, Jeong H-J, Kim S-L, Kim EM, Kim DW, Lim ST, et al. SPIONloaded chitosan-linoleic acid nanoparticles to target hepatocytes. International Journal of Pharmacy. 2009;**371**:163

[133] Hari K, Pichaimani A, Kumpati P. Acridine orange tethered chitosan reduced gold nanoparticles: A dual functional probe for combined photodynamic and photothermal therapy. RSC Advance. 2013;**3**:20471-20479

[134] Salehizadeh H, Hekmatian E, Sadeghi M, Kennedy K. Synthesis and characterization of core-shell Fe3O4-gold-chitosan nanostructure. Journal of Nanbiotechnology. 2012;**10**(1):3

[135] Ahmad M, Manzoor K, Ikram S. 2-Chitosan based nanocomposites for drug, gene delivery, and bioimaging applications. In: Inamuddin AAM, Mohammad A, editors. Applications of Nanocomposite Materials in Drug Delivery. Woodhead Publishing; 2018. pp. 27-38. DOI: 10.1016/B978-0-12-813741-3.00002-9

[136] Ronkainen NJ, Halsall HB, Heineman WR. Electrochemical biosensors. Chemical Society Reviews. 2010;**39**(5):1747-1763

[137] Yao J, Yang M, Duan Y. Chemistry, biology, and medicine of fluorescent nanomaterials and related systems: New insights into biosensing, bioimaging, genomics, diagnostics, and therapy. Chemical Reviews. 2014;**114**(12):6130-6178

[138] Shan C, Yang H, Han D, Zhang Q, Ivaska A, Niu L. Graphene/AuNPs/chitosan nanocomposites film for glucose biosensing. Biosensors & Bioelectronics. 2010;**25**(5):1070-1074

[139] Li G, Liao JM, Hu GQ, Ma NZ, Wu PJ. Study of carbon nanotube modified biosensor for monitoring total cholesterol in blood. Biosensors & Bioelectronics. 2005;**20**(10):2140-2144

[140] Singh S, Solanki PR, Pandey MK, Malhotra BD. Cholesterol biosensor based on cholesterol esterase, cholesterol oxidase and peroxidase immobilized onto conducting polyaniline films. Sensors and Actuators B: Chemical. 2006;**115**(1):534-541

[141] Liu S, Kang M, Yan F, Peng D, Yang Y, He L, et al. Electrochemical DNA biosensor based on microspheres of cuprous oxide and nano-chitosan for Hg (II) detection. Electrochimica Acta. 2015;**160**:64-73

[142] Yang L, Ren X, Tang F, Zhang L. A practical glucose biosensor based on Fe(3)O(4) nanoparticles and chitosan/nafion composite film. Biosensors & Bioelectronics. 2009;**25**(4):889-895

[143] Singh J, Srivastava M, Kalita P, Malhotra BD. A novel ternary NiFe2O4/CuO/FeO-chitosan nanocomposite as a cholesterol biosensor. Process Biochemistry. 2012;**47**(12):2189-2198

[144] Kaushik A, Solanki PR, Pandey MK, Ahmad S, Malhotra BD. Cerium oxide-chitosan based nanobiocomposite for food borne mycotoxin detection. Applied Physics Letters. 2009;**95**(17):173703

[145] Cui HF, Wu WW, Li MM, Song X, Lv Y, Zhang TT. A highly stable acetylcholinesterase biosensor based on chitosan-TiO_2-graphene nanocomposites for detection of organophosphate pesticides. Biosensors & Bioelectronics. 2018;**99**:223-229

[146] Murugesan S, Scheibel T. Chitosan-based nanocomposites for medical applications. Journal of Polymer Science. 2021;**59**:1610-1642

[147] Esmaeili Y, Khavani M, Bigham A, Sanati A, Bidram E, Shariati L, et al. Mesoporous silica@chitosan@ gold nanoparticles as "on/off" optical biosensor and pH-sensitive theranostic platform against cancer. International Journal of Biological Macromolecules. 2022;**202**:241-255

[148] Boubezari I, Bessueille F, Bonhomme A, Raimondi G, Zazoua A, Errachid A, et al. Laccase-based biosensor encapsulated in a galactomannan-chitosan composite for the evaluation of phenolic compounds. Biosensors (Basel). 2020;**10**(6):70

[149] Hroncekova S, Bertok T, Hires M, Jane E, Lorencova L, Vikartovska A, et al. Ultrasensitive Ti3C2TX MXene/chitosan nanocomposite-based amperometric biosensor for detection of potential prostate cancer marker in urine samples. Processes (Basel). 2020;**8**(5):580

[150] Dabhade A, Jayaraman S, Paramasivan B. Development of glucose oxidase-chitosan immobilized paper biosensor using screen-printed electrode for amperometric detection of Cr(VI) in water. 3 Biotech. 2021;**11**(4):183

[151] Gupta R, Goddard NJ. A study of diffraction-based chitosan leaky waveguide (LW) biosensors. The Analyst. 2021;**146**(15):4964-4971

[152] Li H, Long M, Su H, Tan L, Shi X, Du Y, et al. Carboxymethyl chitosan assembled piezoelectric biosensor for rapid and label-free quantification of immunoglobulin Y. Carbohydrate Polymers. 2022;**290**:119482

[153] Li W, Zhao Z, Yang W, Su Q, Na C, Zhang X, et al. Immobilization of bovine hemoglobin on Au nanoparticles/ MoS2 nanosheets—Chitosan modified screen-printed electrode as chlorpyrifos biosensor. Enzyme and Microbial Technology. 2022;**154**:109959

[154] Xu L, Liang X, You L, Yang Y, Fen G, Gao Y, et al. Temperature-sensitive poly(N-isopropylacrylamide)-chitosan hydrogel for fluorescence sensors in living cells and its antibacterial application. International Journal of Biological Macromolecules. 2021;**189**:316-323

Chapter 6

Influence of Loading Nanoclay on Properties of the Polymer-based Composite

Oluwatoyin Joseph Gbadeyan, Zikhona Linganiso and Nirmala Deenadayalu

Abstract

The recent techniques for improving polymer-based composite properties using nanoclay infusion have been reviewed in this chapter. The recent progress in the printing of thermoplastic composite infused with different sizes of particles was reviewed. The processing of infusing clay into natural fiber and recent advancements in the printing of thermoplastic composite infused with nanoclays at different loading ratios also was discussed. Valid information on different apparatuses for determining mechanical properties, temperature dependence storage modulus, and tan α of the developed materials were provided. The loading effect of clay on the mechanical properties, temperature dependence storage modulus, and tan α of composite and nanocomposites was reviewed. Specific emphasis on printed nanocomposite application in gears and related engineering fields is considered. The innovative scope of infusing nanoclay for developing composite with improved mechanical properties, temperature dependence storage modulus, and tan α was discussed. Similarly, the application of clay-reinforced composite with the revolutionary scope of infusing nanoclay for different applications was suggested.

Keywords: clay, nanocomposite, composite, 3D printing, infusion techniques

1. Introduction

Polymers are the most flexible materials for several applications, including but not limited to automotive, construction, biomedical, packaging, aerospace, electronics, and packaging, to mention but a few. This material may be either synthetic binders or greener resin systems, which could be either thermosetting or thermoplastic polymer mainly used to provide shape stability and rigidity to composite materials [1]. These materials are widely used for several applications due to their inherent mechanical, thermal, electrical, optical, and tribological properties. It is well known that polymer matrix plays a vital role in reinforced polymer composites. The primary functions of the polymer matrix are to protect reinforcement, provide rigidity, and hold filler or reinforcement orientation in a specific configuration [2, 3]. Despite the acceptability of polymetric material for various applications, these materials have some constraints, reversing some of the above-mentioned properties. These limitations, such as low thermal conductivity and load-bearing capacity with a higher tendency to creep, poor thermal stability, better heat dissipation, and a very high coefficient of thermal expansion, provoke the clearance problem.

2. Effect of incorporating filler materials on polymer-based composite materials

Several efforts have been explored toward solving the challenges mentioned above—loading several reinforcements or nanofiller into polymetric material to produce a polymer-based composite with improved properties [4–11]. Dittanet and Pearson confirmed that incorporating filler is a viable way to improve the thermal, physical, and chemical of polymer-based composite or hybrid. Furthermore, adding filler/s has been adopted to reduce the cost of the expensive polymetric matrix for composite material development [4–8].

Several materials with required reinforcement have been manufactured into different sizes and used as fillers in polymer composites material with improved properties [9, 12–15]. Fillers and reinforcement produced from SiO_2, TiO_2, and carbon-based materials have been widely explored [8, 12, 14, 16–18]. Among the material used for reinforcement or fillers, carbon-based materials and nanoclay are commonly used. Nanoclay is commonly used as filler or reinforcement in pulp and papers, paints, and polymer-based composite industries. Over many decades, montmorillonite (MMT) and kaolinite with a high concentration of $CaCO_3$, commonly referred to as nanoclay, have been obtained from rock and synthesized using either gas pressure blasting or explosion method [19–21]. This filler is widely accepted and used for its reinforcement potential, availability, and low cost [11, 22, 23]. Furthermore, the development and application of nanoparticle-reinforced polymer-based composite have significantly increased over the decades. This increase in usage could be attributed to their availability, cost-effectiveness, easy processability, improved strength, stiffness, and lightweight [24, 25].

Yao and You [26] classified nanoclay-layered mineral silicate as montmorillonite, bentonite, and kaolinite in agreement with their chemical composition and morphology. They confirmed that montmorillonite is mostly used in industry and research. Mohan and Kanny confirmed the extended use of montmorillonite [14] as they confirmed that montmorillonite nanoparticle loading is a viable way of improving the composite mechanical, thermal, and tribological properties. However, the concentration and particle sizes of nanoclay have been explored, and it has been confirmed that these two factors substantially affect the inclusion of nanoclay in polymer composite properties.

3. Influence of particle sizes and loading ratio of fillers on polymer composite materials

Authors have extensively investigated the effect of particle sizes and the addition ratio of fillers polymer composite properties [10, 15, 18, 27]. Their results proved that the loading of filler at different particle sizes and volumes often has different effects on composite, most times improving one property and harmfully affecting another in most cases [10, 15, 27]. These factors have been considered, and the positivity aspect has been capitalized, and work around the negative aspect through innovative moves and techniques. These factors led to much research to simplify the procedure for synthesizing and characterizing the fiber/filler-reinforced composite and authenticate the improvement in properties and the mechanism that governs the performance for repeatability.

Yue et al. [15] confirmed that the concentration effect of filler on polymer-based composite by establishing the amount of filler incorporated in the polymer composite strongly determines its thermal properties. Furthermore, Fröhlich et al. [9] and Donnet [23] provided information on the characterization and reinforcement loading effect on polymer-based composite, which provided an understanding of the kind of filler loaded in rubber. Nanoclay particle sizes, irrespective of chemical and morphological properties, proved to have dominated effect on polymer composite where clay is used as reinforcement [10, 28, 29]. Laouchedi et al. [28] investigated the effect of locally produced particle sizes and loading rate on epoxy composite material's physical and mechanical properties. According to this study, the loading of more significant size particles damagingly influences composite properties. However, epoxy composite properties were improved after loading smaller particle sizes and loading of 2%wt. Similarly, several studies results are consistent with what Laouchedi et al. reported, proving that incorporating smaller particle sizes and loading of nanoclay are viable ways to produce composite material with improved properties. The discovery and loading of this filler with suitable loading concentrations and particle sizes have given birth to different composite materials adopted for different applications.

Aside from the dispersion of nanoclay in the matrix to enhance polymer composite, nanoclay has been used as an interfacial treatment agent for natural fiber. This filler treatment eventually resulted in improving natural fiber-reinforced composite. Furthermore, techniques like the infusion of nanoclay of nanoclay-layered mineral silicate on the printed layer to develop gear material have been explored [30, 31], and this chapter will provide information on how these techniques, clay particle sizes, and concentrations affect polymer composite properties.

4. Infusion of nanoclay-layered mineral silicate into natural fiber for improved composite properties

Over the years, natural fibers have been chemically treated to remove wax, surface lignin, and another amorphous phase toward increasing properties [28, 32]. Researchers treat natural fiber to possess required properties that could be an alternative reinforcement to synthetic fiber. These treatment techniques have attracted several types of research resulting in the development of materials for different applications [33]. Mohan and Kanny [13] inched this process by infusing nanoclay into banana fiber through shear-induced force using sodium hydroxide chemical treatment techniques. The effect of the infused nanoparticle on the composite's structural, mechanical, morphological, and thermal properties was investigated and compared with composite reinforced with sodium hydroxide chemically treated fibers [13]. The aim of infusing nanoclay with alkaline treatment was expected to enhance the interfacial adhesion of fiber to increase further the mechanical and thermal properties of banana fiber-reinforced composite. Similarly, this study achieved two aims: remove unwanted hemicellulose, lignin, and amorphous phases of banana fiber and incorporate nanoclay to increase its concentration of nanoparticle in-house cellulose phases. Achieving these aims shows that nanoclay could be incorporated to enhance natural fiber-reinforced biocomposite either by infusing nanoclay into in-house cellulose phases of fiber and infuse matrix or by dispersing it in matrix and infused into fiber mat or produce the composite using casting techniques.

5. Materials and methods

Sodium hydroxide, clay, and banana fiber were the raw materials used for this study.

5.1 Chemical treatment and nanoclay infusion into the banana fiber

Nanoclay was infused into fibers' in-house cellulose phases using shear-induced force treatment in two phases, which are alkaline (NaOH) and NaOH/clay treatment. Before chemical treatment, fiber was extracted from the banana plant, air-dried for a week, and chopped to a 5 cm uniform length. Similarly, removing unwanted phases before the chopped banana fiber achieved alkaline treatment was soaked, starred slowing in acetone for 30 min, and then oven-dried at 80°C for 120 min.

5.2 Fibers chemical treatment (NaOH)

Forty grams of dried chopped fiber, 40 grams of sodium hydroxide, and 600 ml of water were measured simultaneously for this process. Then, sodium hydroxide was added to distilled water and stirred at 500 rpm under a controlled temperature of 80°C for 15 min. Afterward, 40 grams of dried chopped fiber were added to the sodium hydroxide solution and stirred for another for 4 h at the same temperature (80°C). Later, the treated fibers were removed from the sodium hydroxide solution and softly washed using running water to remove retaining sodium hydroxide solution or unwanted phases in the fiber to ensure purity. The fiber was then dried at 80°C for 2 h and later used to develop banana fiber-reinforced composite and hybrid.

5.3 Sodium hydroxide/nanoclay treatment

Montmorillonite (MMT) nanoclay was a shear-induced force into banana fiber using the alkane treatment technique. Forty grams of sodium hydroxide were poured into 600 ml of water and stirred at 500 rpm at a conditional temperature of 80°C for 15 min. After which, 20 grams of Na^+ montmorillonite were put into the sodium hydroxide solution and stirred together for 30 min to dissolve the clay particle in the solution. Chopped banana fiber (20 grams) was added to montmorillonite/sodium hydroxide solution and mixed for another 4 h at 750 rpm under the exact temperature of 80°C. The high stirring speed generates a high shear force generated, which enhance the dispersion of clay particle into fibers. Later, the treated fibers were removed from the montmorillonite/sodium hydroxide solution and softly washed using running water to remove retaining montmorillonite/sodium hydroxide solution or unwanted phases in the fiber to ensure purity. The fiber was then dried at 80°C for 2 h and later used to develop banana fiber-reinforced composite and hybrid.

5.4 Composite development

The treated and untreated fiber reinforced epoxy resin using conventional resin casting techniques. This process was employed to develop both untreated and treated banana fiber-reinforced composite. The fibers were chopped into different lengths between 30 and 50mm, and fiber with a specific length was incorporated as a reinforcement in the matrix (epoxy resin). This process was aimed to determine the required length and concentration of the fibers treated and untreated for these series and also to identify reinforcement suitability and influence of each fiber length on epoxy resin properties. The determination of critical length is a prerequisite to fiber concentrations in epoxy resin, meaning the fiber volume was committed only when the critical length was discovered.

The conventional resin casting techniques were conducted in two phases. The first phase was mixing fiber and resin, and the second entailed casting fiber/resin in a mold cavity. The first phase was achieved by measuring 100 grams of epoxy resin into a beaker and heating it to 80°C. Then, a specific concentration with fiber length was added and stirred for 1 hour with a magnetic stirrer at 500 rpm, still at a temperature of 80°C. This blend was removed from the stirrer, allowing it to cool down at ambient temperature for 30 min. A 10:30 mixing ratio of epoxy resin and catalyst, which had the 30%, was adopted. The catalyst mixing ratio of 30 was afterward added to the fiber-resin blend and stirred till homogeneousness was achieved. As mentioned, casting the fiber-resin blend in the mold cavity is the second. This process was achieved by pouring the fiber-resin catalyst into the top open-ended Perspex mold sheet gapped with a 3-mm rubber gasket on three sides. The composite cavity was quickly removed after curing by applying wax on the inner part of the mold before pouring fiber-resin catalyst. The casted banana fiber-reinforced epoxy resin composite was removed after 48 h, and its properties were investigated after 7 days of the initial casting—the effect of banana fiber distribution during casting on epoxy resin thermomechanical properties.

5.5 Testing

The thermal behaviors of the fiber and composite were investigated on a thermogravimetry (SDT Q600 model). The heating profile was obtained at a 10 °C/min heating rate under a dry nitrogen gas flow at 100 mL/min from 0°C to 600°C. Temperature dependence mechanical properties such as storage modulus and tan α were measured on a dynamic mechanical analyzer at a frequency of 10 Hz in a three-point bending mode. This investigation was carried out using the TA equipment (Q800) from room temperature to 100°C under atmospheric conditions. Resistance to pulling stress (tensile strength) and fiber and composite residual strength was investigated using an MTS-UTM machine with a 1 KN load cell. Short beam investigation was used to determine the shear properties of the untreated and treated fiber-reinforced epoxy composite series. This test was carried out in agreement with ASTM 2344-84 standard using a crosshead speed of 1 mm/min. The tensile test was carried out following ASTM D 3039 using an operating speed of 1 mm/min with a 1 KN load cell. The samples investigated were three cubical blocks of 1 cm by 1cm by 1cm cut from banana fiber (treated and untreated) reinforced composite. A fiber pull-out investigation was carried out on samples prepared with 10 cm fiber length entrenched in 3 mm depth in cured epoxy resin, which was subjected to a constant load of 1 kN until a failure occurred. Fiber pull-out strength was determined using the failure ratio on a load-displacement curve. Samples were stressed for both pull-out and tensile strength until the coefficient of variance (CV) was ≤16%.

5.5.1 Interfacial and tensile strength

The tensile strength of the untreated and treated banana fiber shown in **Figure 1** depicts that the infusing of nanoclay and structural alterations due to chemical treatment has a strong positive effect on the tensile properties of the fiber. The tensile strength, stiffness, and elongation at break values shown in **Figures 1–3** vary among untreated and treated fiber. The untreated fiber (UTBF) exhibited a tensile strength of 602 MPa and stiffness of 17 GPa with 4.3% elongation at break. An increase in tensile strength and stiffness after banana fiber was treated was with sodium hydroxide only (NTBF). The composite series exhibited tensile strength and stiffness of 713 MPa and 22 GPa. This performance depicts the effect of the chemical treating banana fiber using sodium hydroxide. A further increase in strength and stiffness was observed after the banana fiber was treated with a combination of sodium hydroxide and clay.

Composite reinforced with banana fiber treated with sodium hydroxide, and clay (N&CTBF) exhibited a tensile modulus of 43 GPa, approximately threefold of

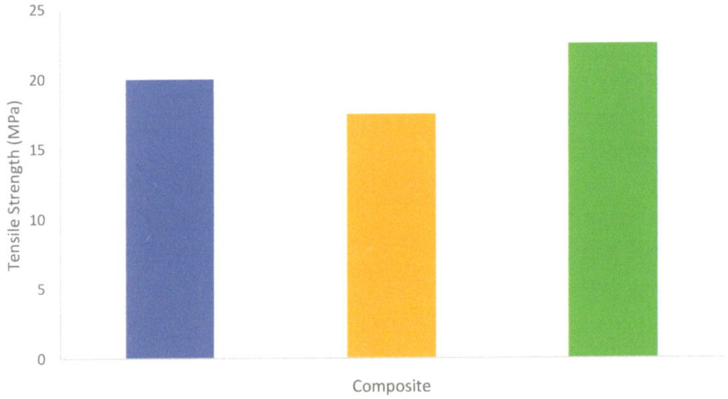

Figure 1.
Tensile strength of composite series.

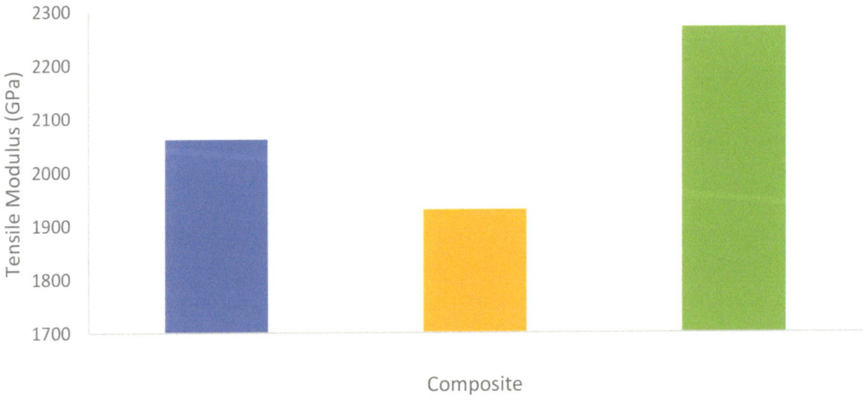

Figure 2.
Tensile modulus of composite series.

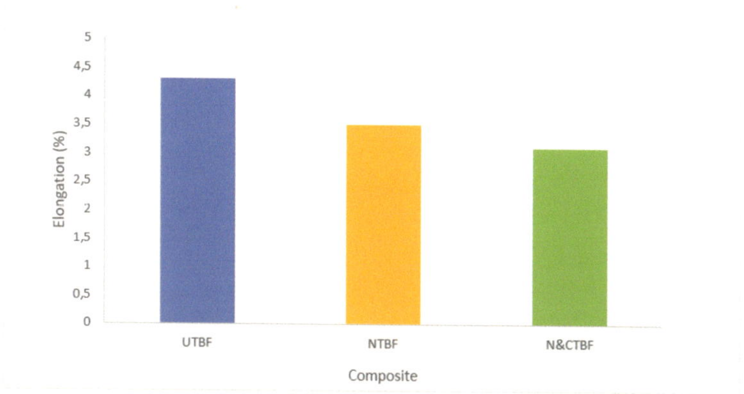

Figure 3.
Elongation at the break for composite series.

untreated fiber stiffness, and tensile strength of 918 MPa, 51% higher than what is obtained with untreated fiber. Removing weak strength phases such as wax, lignin, hemicellulose, and other impurities from the fiber could be the primary reason for the observed tensile strength and stiffness.

Furthermore, the reinforcement effect of the infuse nanoclay is another reason for a significant improvement in tensile strength and stiffness. The increase in tensile stiffness and strength may be attributed to interfacial bonding between the matrix and the fiber due to the presence of nanoclay in the composite.

However, a linear reduction in elongation at break was seen. Composite with untreated fiber exhibited the highest elongation value of 4.3%, followed by composite reinforced with banana fiber treated with sodium hydroxide only, which is 3.5%, and composite filled with banana fiber treated with a combination of sodium hydroxide and clay exhibited the lowest elongation of 3.1%. This decrease in elongation value may be attributed to the absence of banana fiber's waxy and amorphous phases, which were removed during chemical treatment.

5.5.2 Temperature-dependent mechanical properties

Figures 4 and **5** show the effect of the chemical treatment given to banana fiber on dynamic mechanical properties such as storage modulus and tan α. It was observed that the composite reinforced with untreated banana fiber exhibited high storage modulus of 6380 MPa at room temperature and declined with a corresponding increase in temperature. Fibers treated with sodium hydroxide solution and montmorillonite/sodium hydroxide filled composite exhibited 7553 and 8328 MPa at room temperature. This result depicts that giving fiber chemical treatment using a combination of montmorillonite and sodium hydroxide is a viable way to improve temperature-dependent storage modulus. The composite with banana fiber treated using montmorillonite/sodium hydroxide exhibited a temperature-dependent modulus at room temperature, 32% higher than the composite with untreated banana fiber.

The tan α curves shown in **Figure 5** provided details on the composite series' phase transformation, damping features, and interfacial strength. An increase in tan α value with a corresponding increase in temperature is till attaining a maximum point, followed by a decline in temperature. The temperature where tan α reaches a maximum

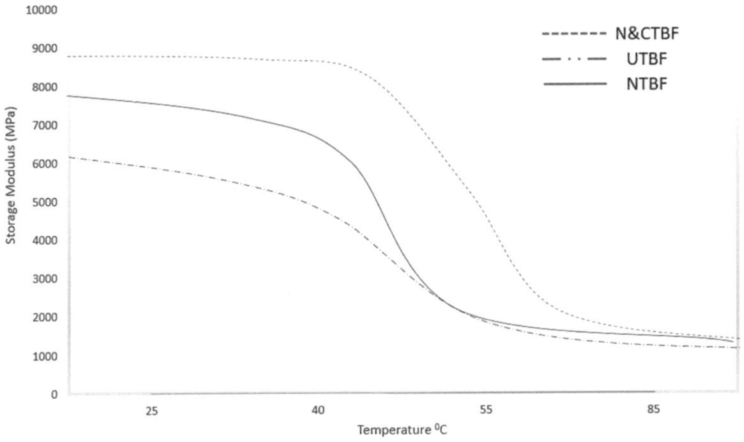

Figure 4.
Temperature dependence storage module of composite series.

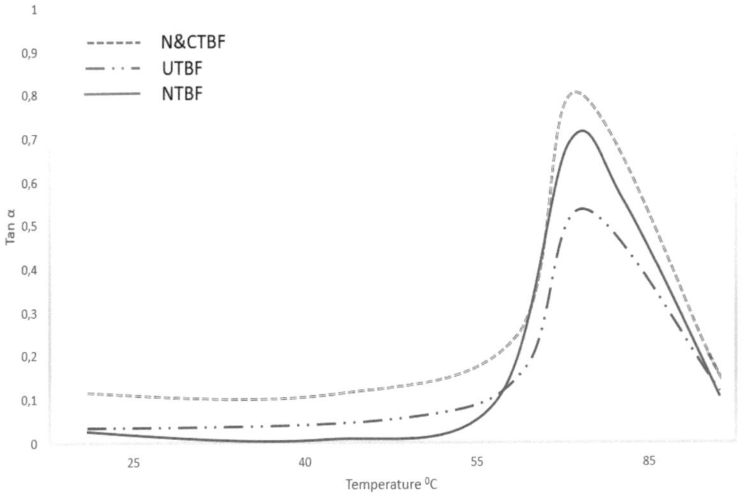

Figure 5.
Temperature dependence tan α of composite series.

value, referred to as the glass transition temperature of the polymer, was identified and recorded. At this glass transition temperature, the polymer changes from an amorphous to a rubbery state under increased temperature.

Composite with untreated fiber, treated with sodium hydroxide, and montmorillonite/sodium hydroxide exhibited glass transition at 69°C, 69°C, and 72°C, respectively. This output implied that composite filled with fiber treated with montmorillonite/sodium hydroxide, and this performance might be attributed to the presence of nanoclay in banana fiber. Nanoclay is known for its good thermal properties and ability to delay polymer deformation from a glassy region into a rubbery region [30, 31, 34]. Thus, infusion and dispersion of nanoclay into fiber may have reduced the polymer deformation leading to increased glass transition of composite with fiber treated with montmorillonite/sodium hydroxide. This result trend corresponds with what is experienced in the storage modulus.

These findings proved that the advanced way of infusing clay into natural fiber (banana fiber) could be used to produce composite materials with improved properties for different applications. Going by this technique, clay concentration for producing composite using conventional methods such as resin casting method and vacuum infusion may reduce to minimal, eventually affecting composite production cost positively. We employ infusing nanoclay on acrylonitrile butadiene styrene printer layers using a 3D printer. This technique was used to develop a gear material and determine printed material's mechanical, thermal, and tribological properties. This idea was motivated by the challenging process of gear using conventional techniques, material selection, and production parameters [31]. The research aimed to reduce the time used for producing gear, selecting nanoclay to enhance the interfacial bonding acrylonitrile butadiene styrene printer layers toward developing more robust structures using relatively simple 3D printing processes.

6. Experimental details

6.1 The 3D printing process of nanoclay-infused acrylonitrile butadiene styrene composite

A modified nanoclay and acrylonitrile butadiene styrene filament were explored for the development of gear material. The process was achieved using 3D printing and infusion techniques, which were accomplished in four steps. Drafting gear samples using a computer-aided program was the first step. A gear was designed and drawn on a Solid Edge 2019 software by creating a specific dimension of the gear for creating an ISO metric file as shown in **Figure 6** and was saved as Standard Tessellation.

This file was then exported to UP studio for finalizing the printing parameters. Afterward, acrylonitrile butadiene styrene filament was infilled into a 3D printer set at 99% with no support. Simultaneously, nanoclay was dissolved in acetone using a 10:1 mixing ratio to facilitate uniform dispersion of particles. Subsequently, a layer of acrylonitrile butadiene styrene was printed, then a premeasured acetone/nanoclay solution was applied to the first printed surface (**Figure 7**).

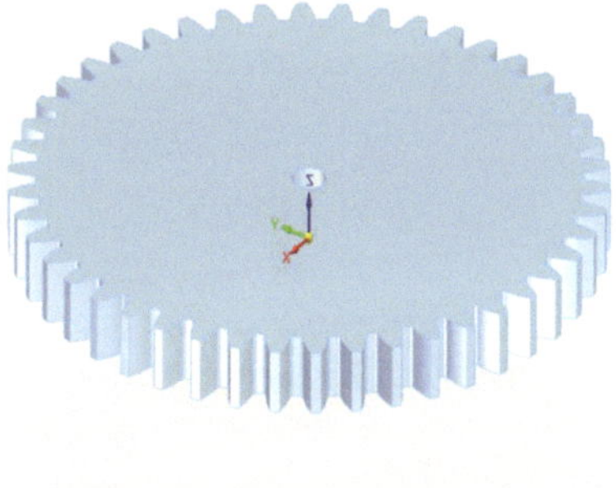

Figure 6.
Gear sample drafted on Solid Edge 2019.

Figure 7.
Acetone/nanoclay solution-coated acrylonitrile butadiene styrene layer [31].

This process was repeated for six layers of printed nanoclay-infused acrylonitrile butadiene styrene-infused composite at equal spacing. All printed layers were coated uniformly to prevent vacuums in the 3D-printed composite. After the printing, the composite was removed and cured under ambient temperature. The effect of nanoclay infusing on the mechanical properties of 3D-printed acrylonitrile butadiene styrene was determined by varying loading from 0.5 to 5%. These investigations were conducted after 7 days.

6.2 Investigation procedures

3D-printed acrylonitrile butadiene styrene infused with nanoclay of different percentage loading resistance to impact, flexural, and tensile properties was investigated to determine the strength and stiffness of the printed nanocomposite. Tensile was carried out on a universal testing machine fitted with a 30 N load cell using ASTM 3039 test standard. Impact investigation was conducted according to ASTM D6110-10 using a Hounsfield Balance Impact Tester. For each investigation, five samples were investigated, and the average value was reported.

7. Results and discussion

Mechanical properties such as tensile and impact properties of 3D-printed nanocomposite were shown in **Figures 8–10**. **Figures 8** and **9** illustrate the tensile strength and modulus of neat acrylonitrile butadiene styrene and 3D-printed nanocomposite infused with different percentages of nanoclay. Although the tensile strength and stiffness vary with different loading percentages, the infusion of nanoclay increased both tensile strength and stiffness of neat acrylonitrile butadiene styrene, irrespective of loading weight percentages.

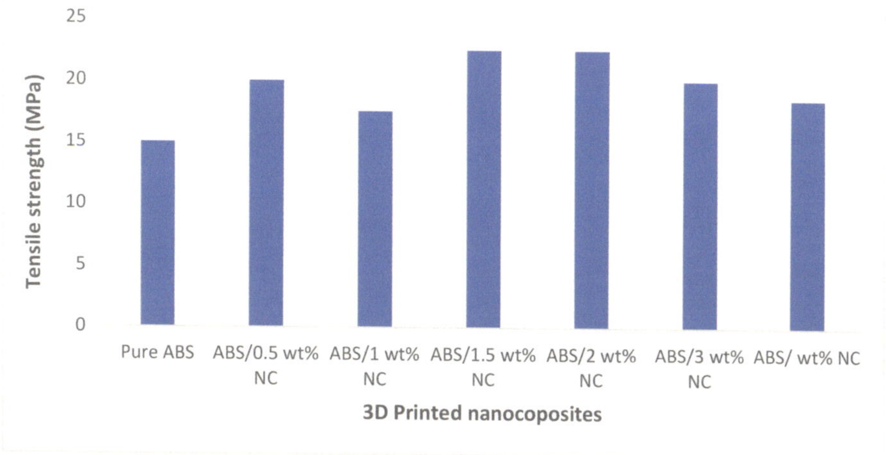

Figure 8.
Tensile strength of acrylonitrile butadiene styrene and nanoclay infused nanocomposite.

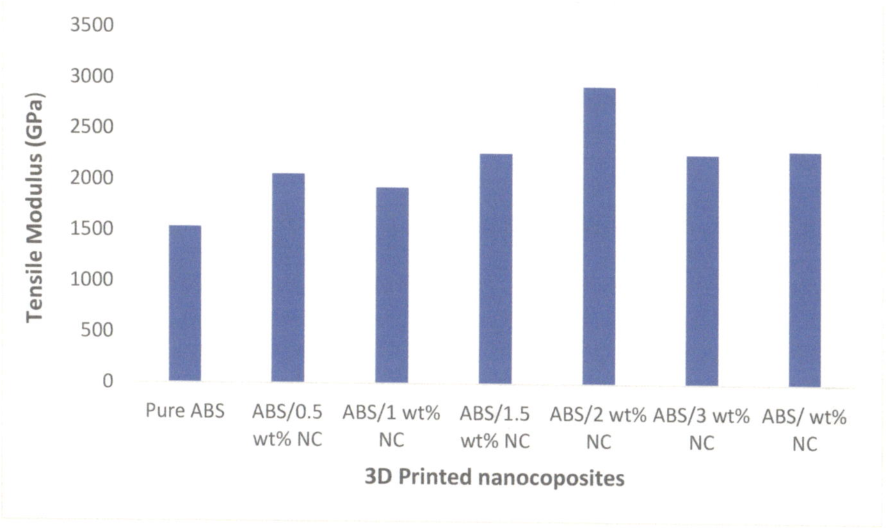

Figure 9.
Tensile stiffness of acrylonitrile butadiene styrene and nanoclay-infused nanocomposite.

However, superior tensile strength and modulus were achieved when 2%.wt nanoclay was infused. The addition of 2%.wt. improved tensile strength by 50% and increased acrylonitrile butadiene styrene tensile modulus by 90%. The apparent increase in tensile properties can be attributed to the presence of nanoclay serving as a reinforcing agent in the nanoclay-infused nanocomposite [35–37]. This performance depicts the strong positive effect on the pulling resistance of acrylonitrile butadiene styrene. Furthermore, the tensile strength and stiffness showed by acrylonitrile butadiene styrene solely depend on the amount of nanoclay infused as printed nano-composite tensile properties increased with nanoclay loading. Infusion of nanoclay on every layer of printed acrylonitrile butadiene styrene may have increased interfacial bonding, increasing resistance to pulling stresses.

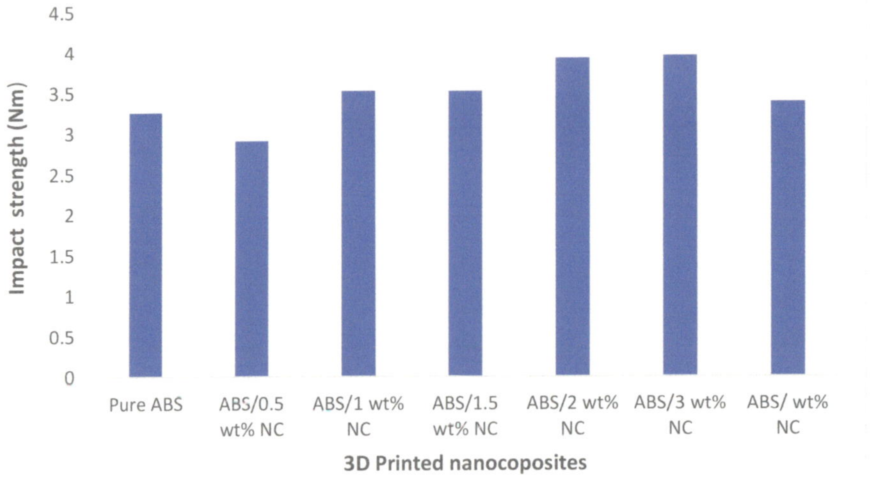

Figure 10.
Impact strength of acrylonitrile butadiene styrene and nanoclay-infused nanocomposite.

Besides, nanoclays are known for their stiffness, much stiffer than neat acrylonitrile butadiene styrene. This stiffness may be attributed to better resistance to stress yielding acrylonitrile butadiene styrene when subjected to pulling stress, resulting in nanocomposite stiffness improvement [35–37].

7.1 3D-printed nanocomposite resistance to impact stress

Figure 10 shows resistance to impact stress of printed acrylonitrile butadiene styrene and nanoclay-infused nanocomposite. Similar to the trend observed in **Figures 8** and **9**, the impact strength of printed nanocomposite varies with nanoclay loading. Infusing nanoclay enhances the impact property of printed acrylonitrile butadiene styrene. 3D-printed nanocomposite infused with 3 wt.% exhibited superior impact resistance with 20% higher than neat acrylonitrile butadiene styrene. However, trivial increased impact strength was observed with a corresponding increase of nanoclay loading from 2 wt.% (3.93) to 3 wt.% (3.96). This resistance to sock performance of 3D-printed nanoclay-infused nanocomposite may be attributed to the critical loading efficiency of nanoclay at lower concentrations. Besides, the interfacial bonding introduced by nanoclay infused on the acrylonitrile butadiene styrene printed layers may be attributed to the improvement in impact strength observed.

The infusion of nanoclay strengthens the bonding at the interface of each printed layer, improving the energy absorption capability and impact strength. A sharp drop in impact resistance at loading higher than 3 wt.% of nanoclay may be attributed to particle agglomeration, causing focal stress areas that exhibited poor resistance to impact stress.

8. Conclusion

This chapter reviews the influence of recent clay infusion on mechanical and temperature dependence storage modulus and tan α of polymer-based composite and nanocomposite. Progress in techniques for developing clay-reinforced polymer-based composite and hybrid nanocomposite was examined. The infusing of clay into a naturally sourced banana fiber later used in the development of composite was discussed. Infusion of nanoclay into printed layers of acrylonitrile butadiene styrene was discussed. The effect of infused nanoclay on polymer-based composite properties was reported. It provides information on pieces of equipment used for developing polymer-based composite and nanocomposite and properties evaluation. This review confirmed that infusing clay into the banana fiber helps in tailoring and improving composite properties to achieve desired properties, predominantly mechanical and temperature dependence storage modulus and tan α. It also suggested the developed materials for the application of interior parts in automobile industries. Furthermore, the infusion of nanoclay on printed layers of acrylonitrile butadiene styrene significantly improves the mechanical properties of the printed nanocomposite. 3D-printed acrylonitrile butadiene styrene-infused nanoclay with improved mechanical properties suggested materials for a gear development application.

Author details

Oluwatoyin Joseph Gbadeyan[1,2*], Zikhona Linganiso[1,2,3]
and Nirmala Deenadayalu[1]

1 Department of Chemistry, Durban University of Technology,
Republic of South Africa

2 Faculty of Engineering and Built Environment, Green Engineering Research Focus Area, Durban University of Technology, Durban, South Africa

3 Faculty of Applied Sciences, Durban University of Technology, Durban, South Africa

*Address all correspondence to: oluwatoyinG@dut.ac.za

References

[1] Paul V. Synthesis and characterization of a biocomposite derived from banana plants (Musa cavendish) [Doctor of Philosophy]. Chemical Durban University of Technology; 2015

[2] Bijwe J, Majumdar N, Satapathy B. Influence of modified phenolic resins on the fade and recovery behavior of friction materials. Wear. 2005;**259**(7-12):1068-1078

[3] Cai P, Wang Y, Wang T, Wang Q. Effect of resins on thermal, mechanical and tribological properties of friction materials. Tribology International. 2015;**87**:1-10

[4] Gbadeyan OJ. Low friction hybrid nanocomposite material for brake pad application. 2017

[5] Gbadeyan OJ, Kanny K, Mohan TP. Influence of the multi-walled carbon nanotube and short carbon fibre composition on tribological properties of epoxy composites. Tribology – Materials, Surfaces and & Interfaces. 2017;**11**(2):59-65

[6] Saheb DN, Jog JP. Natural fiber polymer composites: A review. Advances in Polymer Technology. 1999;**18**(4):351-363

[7] Sanjay MR, Madhu P, Jawaid M, Senthamaraikannan P, Senthil S, Pradeep S. Characterization and properties of natural fiber polymer composites: A comprehensive review. Journal of Cleaner Production. 2018;**172**:566-581

[8] Vijayan P, Puglia D, Al-Maadeed MASA, Kenny JM, Thomas S. Elastomer/thermoplastic modified epoxy nanocomposites: The hybrid effect of 'micro' and 'nano' scale. Materials Science and Engineering. 2017;**116**:1-29

[9] Fröhlich J, Niedermeier W, Luginsland HD. The effect of filler–filler and filler–elastomer interaction on rubber reinforcement. Composites Part A: Applied Science and Manufacturing. 2005;**36**(4):449-460

[10] Katritzky AR et al. Effect of filler loading on the mechanical properties of crosslinked 1, 2, 3-triazole polymers. Journal of Applied Polymer Science. 2010;**118**(1):121-127

[11] Miyagawa H, Jurek RJ, Mohanty AK, Misra M, Drzal LT. Biobased epoxy/clay nanocomposites as a new matrix for CFRP. Composites Part A: Applied Science and Manufacturing. 2006;**37**(1):54-62

[12] Dizaj SM, Barzegar-Jalali M, Zarrintan MH, Adibkia K, Lotfipour F. Calcium carbonate nanoparticles; potential in bone and tooth disorders. Pharmaceutical Sciences. 2015;**20**(4):175

[13] Mohan T, Kanny K. Nanoclay infused banana fiber and its effects on mechanical and thermal properties of composites. Journal of Composite Materials. 2016;**50**(9):1261-1276

[14] Onuegbu GC, Igwe IO. The effects of filler contents and particle sizes on the mechanical and end-use properties of snail shell powder filled polypropylene. Materials Sciences and Applications. 2011;**2**(7):810

[15] Yue Y, Zhang H, Zhang Z, Chen Y. Polymer–filler interaction of fumed silica filled polydimethylsiloxane investigated by bound rubber. Composites Science and Technology. 2013;**86**:1-8

[16] Deogonda P, Chalwa VN. Mechanical property of glass fiber reinforcement epoxy composites. International Journal of Scientific Engineering and Research (IJSER). 2013;**1**(4):2347-3878

[17] Müller CM, Laurindo JB, Yamashita F. Composites of thermoplastic starch and nanoclays produced by extrusion and thermopressing. Carbohydrate Polymers. 2012;**89**(2):504-510

[18] Joseph GO, Adali S, Bright G, Sithole B. Nanofiller/natural fiber filled polymer hybrid composite: A review. Journal of Engineering Science & Technology Review. 2021;**14**(5)

[19] Balachandran M, Devanathan S, Muraleekrishnan R, Bhagawan S. Optimizing properties of nanoclay–nitrile rubber (NBR) composites using face centred central composite design. Materials & Design. 2012;**35**:854-862

[20] Balachandran M, Bhagawan S. Mechanical, thermal and transport properties of nitrile rubber (NBR)—nanoclay composites. Journal of Polymer Research. 2012;**19**:9809

[21] Senatov F, Kuznetsov D, Kaloshkin S, Cherdyntsev V. Obtaining nanopowders of metal oxides from salts by means of mechanochemical synthesis. Chemistry for Sustainable Development. 2009;**17**(6):631-636

[22] Gbadeyan OJ, Adali S, Bright G, Sithole B, Omojoola A. Studies on the mechanical and absorption properties of achatina fulica snail and eggshells reinforced composite materials. Composite Structures. 2020;**2020**:112043

[23] Donnet J-B. Black and white fillers and tire compound. Rubber Chemistry and Technology. 1998;**71**(3):323-341

[24] Ippolito F, Hübner G, Claypole T, Gane P. Calcium carbonate as functional filler in polyamide 12-manipulation of the thermal and mechanical properties. Processes. 2021;**9**(6):937

[25] Lee Y, Kim Y, Kim SR, Shin DG, Oh SC, Kwon WT. Size effect of CaCO3 filler on the mechanical properties of SMC composites. In: Defect and Diffusion Forum. Switzerland: Trans Tech Publ; 2015. pp. 244-248

[26] Yao H, You Z. Nanoclay modified asphalt. In: Innovative Developments of Advanced Multifunctional Nanocomposites in Civil and Structural Engineering. USA: Elsevier; 2016. pp. 183-216

[27] Gbadeyan O, Kanny K, Pandurangan MJJTT. Tribological, mechanical, and microstructural of multiwalled carbon nanotubes/short carbon fiber epoxy composites. Journal of Tribology. 2018;**140**(2):022002-022008

[28] Laouchedi D, Bezzazi B, Aribi C. Elaboration and characterization of composite material based on epoxy resin and clay fillers. Journal of Applied Research and Technology. 2017;**15**(2):190-204

[29] Friedrich K, Zhang Z, Schlarb AK. Effects of various fillers on the sliding wear of polymer composites. Composites Science and Technology. 2005;**65**(15-16):2329-2343

[30] Gbadeyan OJ, Kanny K, Mohan T. Tribological properties of layered silicate nanoparticle filled acrylonitrile butadiene styrene (ABS) nanocomposite produced using 3D printing. Polymer-Plastics Technology and Materials. 2022:1-12

[31] Gbadeyan OJ, Mohan T, Kanny K. Processing and characterization of 3D-printed nanoclay/acrylonitrile

butadiene styrene (abs) nanocomposite gear. The International Journal of Advanced Manufacturing Technology. 2020;**109**(3):619-627

[32] Gbadeyan OJ, Adali S, Bright G, Sithole B. Comparative reinforcement effect of Achatina fulica snail shell nanoparticles, montmorillonite, and kaolinite nanoclay on the mechanical and physical properties of greenpoxy biocomposite. Polymers. 2022;**14**(3):365

[33] Mårtensson P. powerRibs™ enables ocean plastic in automotive interior parts. 2019. Available from: http://www.bcomp.ch/en/news/bcomp-enables-upcycling-of-ocean-plastic-for-automotive-interior-parts-in-volvo-cars-recycled-plastics-demonstrator-vehicle. [Accessed January 18, 2019]

[34] Gbadeyan OJ, Mohan TP, Kanny K. Tribological properties of 3D printed polymer composites-based friction materials. In: Jena H, Katiyar JK, Patnaik A, editors. Tribology of Polymer and Polymer Composites for Industry 4.0. Singapore: Springer Singapore; 2021. pp. 161-191

[35] Shen L, Phang IY, Chen L, Liu T, Zeng K. Nanoindentation and morphological studies on nylon 66 nanocomposites. I. Effect of clay loading. Polymers. 2004;**45**(10):3341

[36] Venkatesh G, Deb A, Karmarkar A, Chauhan SS. Effect of nanoclay content and compatibilizer on viscoelastic properties of montmorillonite/polypropylene nanocomposites. Materials & Design. 2012;**37**:285-291

[37] Bashar M, Sundararaj U, Mertiny P. Microstructure and mechanical properties of epoxy hybrid nanocomposites modified with acrylic tri-block-copolymer and layered-silicate nanoclay. Composites Part A: Applied Science and Manufacturing. 2012;**43**(6):945-954